Green Energy and Technology

Climate change, environmental impact and the limited natural resources urge scientific research and novel technical solutions. The monograph series Green Energy and Technology serves as a publishing platform for scientific and technological approaches to "green"—i.e. environmentally friendly and sustainable—technologies. While a focus lies on energy and power supply, it also covers "green" solutions in industrial engineering and engineering design. Green Energy and Technology addresses researchers, advanced students, technical consultants as well as decision makers in industries and politics. Hence, the level of presentation spans from instructional to highly technical.

Indexed in Scopus.

More information about this series at http://www.springer.com/series/8059

Sirichai Koonaphapdeelert · Pruk Aggarangsi ·
James Moran

Biomethane

Production and Applications

Springer

Sirichai Koonaphapdeelert
Department of Environmental Engineering
Chiang Mai University
Chiang Mai, Thailand

Pruk Aggarangsi
Department of Mechanical Engineering
Chiang Mai University
Chiang Mai, Thailand

James Moran
Department of Mechanical Engineering
Chiang Mai University
Chiang Mai, Thailand

ISSN 1865-3529 ISSN 1865-3537 (electronic)
Green Energy and Technology
ISBN 978-981-13-8306-9 ISBN 978-981-13-8307-6 (eBook)
https://doi.org/10.1007/978-981-13-8307-6

This Springer imprint is published by the registered company Springer Nature Singapore Pte Ltd.
The registered company address is: 152 Beach Road, #21-01/04 Gateway East, Singapore 189721, Singapore

To friends, family, and colleagues who helped, encouraged, inspired, and motivated us.

Foreword

Sustainable development goals drive researchers and technologists to find new methods for green growth as well as reducing environmental degradation. Biomethane is a biological-based renewable gas produced from organic wastes and often considered the most cost-effective sustainable renewable energy. Chiang Mai University has been a major player in biogas research and development for more than 20 years with many successful actual scale projects. I congratulate the authors of this book, and this book is written based on research experiences from Chiang Mai University. This book aims to educate the general public as well as engineers of how biogas and biomethane can be utilized in the most efficient and sustainable way. The book also summarizes many actual projects as case studies which shall benefit a large scope of enthusiastic readers.

By Twarath Sutabutr, Sc.D., Director of the Thai Energy Policy and Planning Office (EPPO) 2015–2018.

In Thailand, natural gas (NG) plays an important role in energy security to meet the growing demand for electricity, heating and cooling, and transportation. Natural gas comes with well-developed infrastructure, pipeline system, gas-filling stations, and LNG facilities. On the other hand, biomethane, which is a methane derived and upgraded from biogas with properties close to natural gas, has been gaining ground in Thailand to become a next-generation fuel to support the transition from fossil fuels to renewables and to achieve the Kingdom's aspiration to reduce greenhouse gas emission and increase energy independency targets.

This book is one of the first to focus on the technical stories of the biogas upgrading process to produce biomethane and applications with a variety of real case studies. The book also illustrates the options and needs for the development of biomethane supply strategies. Further, an overview of expected future development of the biomethane sector is also given. Importantly, the book also remarks some major milestones in the history of sustainable energy development in Thailand with biomethane as a key instrument. I personally believe that biomethane will be one of the most sustainable forms of renewable energy with the smallest global warming footprint and even negative in many cases.

Dr. Zaini Ujang
Former President/Vice-Chancellor
Universiti Teknologi Malaysia
Former Secretary-General
of the Malaysian Energy, Green
Technology and Water Ministry

Preface

This is the first edition of our book, Biomethane: Production and Applications. It is authored by Sirichai Koonaphapdeelert, Pruk Aggarangsi, and James Moran. The present team of authors has endeavored to complete this book for publication by Springer Nature. There are a total of seven chapters covering cleaning, processing, and end-use applications of biomethane. Each chapter is a self-contained unit. It is not necessary to read them all in sequence. If the reader is familiar with biomethane and only interested biomethane use in transportation applications, they may skip ahead to Chap. 4. Chapter 1 introduces this subject. Chapter 2 deals with the cleaning and pretreatment processes necessary to begin the upgrading process from biogas to biomethane. Chapter 3 concerns itself with the upgrading process and the post-processing of biomethane. Chapter 4 begins the introduction of applications for biomethane, starting with transportation applications. Chapter 5 is about heating applications for biomethane, in both domestic and industrial applications. The substitution of LPG with biomethane is discussed in detail. Chapter 6 has small-scale, localized gas grids for biomethane use in the local community as its objective. Chapter 7 concludes with potential future uses and processes for biomethane.

The growing importance to develop non-intermittent sources of renewable energies has led to an interest in biomass-derived sources of energy. Biogas is one such source and has its own advantages mainly that its raw materials come from organic waste products. Biomethane is a purified form of biogas that is chemically identical to methane—a fossil fuel. As it is chemically identical, it can be used as a methane replacement. Often, the obstacles in replacing methane are economic. These factors and others are discussed throughout this book.

The aim of this book is to provide a practical engineering description of techniques, processes, and applications in widespread use and provide design criteria to evaluate specific processes. The end of each chapter contains a case study with practical information for design and construction of gas processing equipment. Most of the subject matter comes from the design of biogas and biogas upgrading plants from agricultural farms across Southeast Asia. The case studies are best suited for the climate and conditions in South and Southeast Asia. For example, unlike in Europe, rural areas in Southeast Asia are not close to natural gas pipelines, and therefore, grid injection of biomethane is often not an option. This book focuses on biogas produced from organic agricultural waste and food-processing plants. Biogas and biomethane from landfill gas, sewage sludge, and municipal waste are not discussed in this book. Biogas from these sources typically contains more impurities that need to be removed before combustion. These impurities require an extra processing step.

Several sections of the book deal with economic issues, including plant construction costs, upgrading costs, operating costs and more. Most of these examples are from Thailand where the costs are in Thai baht. Costs, especially labor cost, will of course be different in other countries. Where possible, the costs are presented in US dollars at a conversion rate of $1 = ฿33. Since this conversion rate may fluctuate on a daily basis, prices presented in dollars should be used only as a guide.

No attempt has been made to define the ownership or patent status of the processes described. Many of the basic patents on the well-known processes have expired. Patents on specific proprietary systems that involve special additives or system flow modifications are not discussed in depth. Technical data here are presented in SI units where possible. In some circumstances, when the original data are presented in English units, then they are left as such. Units for production of biomethane are usually expressed as Nm^3/h, which stands for normal cubic meter per hour under standard conditions of 0°C and 1 atm (101.325 kPa).

Chiang Mai, Thailand

Sirichai Koonaphapdeelert
Pruk Aggarangsi
James Moran

Acknowledgments

The assistance of many individuals who contributed material and suggested improvements is gratefully acknowledged. Thanks are also due to the companies and organizations who graciously provided data and gave permission for reproducing charts and figures. The number of such organizations is too large to permit individual recognition here; however, they are generally identified in the text as the sources of specific data.

The authors wish to thank the researchers and staff at the Energy Research and Development Institute of Nakornping of Chiang Mai University who have continuously been conducting research and development into renewable energies with a focus on biogas and its many applications. Special thanks also go to Mr. Panutat Injaima and Mr. Warut Yuennan for their help with the graphic design. Among the many who have supported and financed this research over the years, special thanks go to the Energy Policy and Planning Office (EPPO) of the Thai Ministry of Energy and to Chiang Mai University.

We would like to acknowledge our partners throughout the years including the Thai Biogas Trade Association, RE Biofuels Co., Chiang Mai Fresh Milk Farm Co., Alensys GmbH, and Evonik Industries AG.

Other significant contributors to this first edition are Christoph Baumann and Chandra Sekaran, our patient editors. Finally, we wish to express gratitude to our families for their support and patience during the preparation of this book.

To all fellow researchers in the field of biogas and biomethane, we wish to extend our deepest gratitude and thanks.

About This Book

This book deals with the processes and applications downstream from biogas production. Biogas is a gas formed in an anaerobic digestion process. Organic matter decomposes when placed in an oxygen-free environment. Biogas can be made from agricultural or general household waste, manure, plants, or food waste; hence, it is considered a renewable energy source. Biogas contains methane, so it can be combusted directly on site. In small-scale plants, this is often the best choice. It also contains other gases such as carbon dioxide and hydrogen sulfide which are not so useful. In medium- or large-scale plants where enough biogas is produced in economic quantities, it becomes necessary to clean and purify the biogas before selling. Biomethane is a gas that results from any process that improves the quality of biogas by reducing the levels of carbon dioxide, hydrogen sulfide, moisture, and other gases. If these gases could be removed entirely, the biomethane that remains is pure methane. The name biomethane refers to the method of production, rather than the gas content. If there is a nearby natural gas pipeline, the biomethane can be injected into the pipeline without any complications. Biomethane plants are normally found in locations with a low population density, in the countryside, close to farms or food-processing plants. In situations where a gas pipeline is not nearby, biomethane downstream applications can include storage, transportation, home heating, industrial use, and small-scale local gas grids. This book discusses each of these applications and lists some of the design criteria and issues with each.

Contents

1 Introduction to Biomethane 1
1.1 Background 1
1.2 Biogas Composition 1
1.2.1 Biogas and Pollution Reduction 3
1.3 Biomethane 4
1.3.1 Biomethane Definition 4
1.3.2 Advantages of Biomethane 5
1.4 Biomethane Use Around the World 7
1.4.1 Sweden 8
1.4.2 Germany 9
1.4.3 USA 11
1.5 Case Study—Chiang Mai University Waste to Energy Plant 11
References 15

2 Biogas Cleaning and Pretreatment 17
2.1 Contaminants 17
2.2 Pretreatment Selection Criteria 20
2.3 H2S Bioscrubber 21
2.4 Biofilter Reactor 22
2.4.1 Selection of H2S Removing Bacteria 25
2.5 Biofilter Components 28
2.5.1 Example of a Biofilter Calculation 34
2.6 Biogas Dehumidification 38
2.6.1 Dehumidification Method 39
2.6.2 Vapor Compression System 39
References 43

3 Biogas Upgrading . . . 47
3.1 Biogas Upgrading to Biomethane . . . 47
3.1.1 Water Scrubbing . . . 47
3.1.2 Pressure Swing Adsorption (PSA) . . . 54
3.1.3 Chemical Scrubbing with Amine Solvent . . . 57
3.1.4 Absorption with Physical Solvents . . . 59
3.1.5 Membrane Separation . . . 61
3.2 Economic Analysis . . . 64
3.2.1 Biogas Cost . . . 64
3.2.2 Cost of Biogas Upgrading . . . 66
3.2.3 Total Biomethane Production Cost . . . 68
3.2.4 Other Costs . . . 70
3.2.5 Technology Comparison . . . 71
3.3 Biomethane Post Treatment . . . 72
3.3.1 Moisture Reduction in Biomethane . . . 72
3.3.2 Odorization . . . 82
3.4 Case Study—Water Scrubbing Plant . . . 83
3.4.1 Design Assumptions . . . 84
3.4.2 General System Considerations . . . 86
3.4.3 Water Scrubbing Design . . . 87
3.4.4 Civil Engineering Design . . . 92
3.4.5 Electrical System Design . . . 92
References . . . 96

4 Biomethane in Transportation Applications . . . 99
4.1 Introduction . . . 99
4.2 Natural Gas Quality for Transportation . . . 102
4.3 Natural Gas Vehicles . . . 103
4.3.1 CNG Development Obstacles . . . 106
4.4 Case Study: CBG versus CNG . . . 107
4.4.1 Results . . . 108
References . . . 109

5 Biomethane in Domestic and Industrial Applications . . . 111
5.1 Liquid Petroleum Gas . . . 111
5.2 Biomethane for Domestic Cooking . . . 112
5.2.1 Domestic Stoves . . . 112
5.3 Fuel Switching Analysis . . . 115
5.3.1 Wobbe Index . . . 115
5.3.2 Flame Stability . . . 117
5.3.3 Nozzle Diameters . . . 118
5.3.4 Pressure Drop . . . 119
5.3.5 Hardware Changes to the Stoves . . . 121

5.4 Energy Output and Efficiency ... 123
5.5 Substituting Biomethane in Industrial Applications ... 124
5.5.1 Biomethane Substitution in Shuttle Kilns ... 125
5.5.2 Theoretical Considerations ... 125
5.5.3 Substitution Results in Kilns ... 127
5.5.4 Emission Testing ... 129
5.6 Economics of Fuel Substitution ... 130
5.7 Case Study—Mee Sin Ceramics ... 133
5.7.1 Mee Sin Ceramics Discussion ... 135
References ... 136

6 Biomethane to Local Gas Grids ... 139
6.1 Biomethane to Local Gas Grids ... 139
6.1.1 Biomethane Purity ... 140
6.1.2 Existing Pipeline Standards ... 143
6.1.3 Class Locations ... 145
6.1.4 Pipe Design ... 146
6.1.5 General Construction Requirements ... 149
6.1.6 Customer Meters and Regulators ... 150
6.1.7 Test Requirements ... 151
6.1.8 Records ... 152
6.1.9 Operations ... 152
6.2 Case Study ... 155
6.2.1 Local Biomethane Grid in Chiang Mai ... 155
6.2.2 Economics of the Biomethane Grid ... 159
References ... 161

7 Biomethane—Future Trends ... 163
7.1 Biogas Production ... 163
7.1.1 Acid-Phase Digesters ... 164
7.1.2 Anaerobic Digestion of Microalgae ... 165
7.1.3 Lignocellulosic Biomass ... 166
7.2 Biogas Upgrading ... 167
7.2.1 Cryogenic Upgrading ... 167
7.2.2 In Situ Methane Enrichment ... 168
7.2.3 Upgrading Biogas with Enzymes ... 168
7.2.4 Future Developments in PSA Technologies ... 168
7.3 Biomethane Applications ... 169
7.3.1 Low-Pressure Biomethane Storage ... 169
7.3.2 Biomethane Liquefaction ... 172
7.3.3 Power to Gas ... 174
7.4 Future Road Map ... 175
References ... 178

Index ... 181

About the Authors

Dr. Sirichai Koonaphapdeelert has a Ph.D. in chemical engineering from Imperial College London and has been teaching and working in the fields of environmental engineering and renewable energy for more than 10 years. During his time as the deputy director of the Energy Research and Development Institute, Nakornping of Chiang Mai University, he was involved in a number of projects related to biogas and biomethane. For instance, the development of water scrubbing and membrane technology for biogas upgrading, the characterization and testing of biomethane in vehicles, the uses of biomethane as the replacement for liquefied petroleum gas in ceramic kilns and the demonstration of biomethane micro-grids for households. Also, he initiated and helped drive a legal framework related to biomethane in Thailand, such as the drafted engineering standard for biomethane gas grid and the regulatory property of biomethane for vehicular uses.

Dr. Pruk Aggarangsi is currently the Director of the Energy Research and Development Institute Nakornping, Chiang Mai University and Assistant Professor in Mechanical Engineering Chiang Mai University, Thailand. He has a wide scope of expertise covering areas such as renewable energy, waste water treatment, numerical and finite element modeling, simulation and analysis, energy balance analysis, robotic system design and mechanical vibration analysis. For the past 12 years, he has participated and in charge of multiple biogas/biomethane projects mostly supported by the Thai Ministry of Energy. Dr. Pruk have had crucial roles in engineering of many biogas constructions projects as well as conducting more than 40 biogas-related technical training workshops. He also plays an important role in driving Chiang Mai University's Smart City-Clean Energy Project aim to initiate sustainable development for the communities around the world.

Dr. James Moran received a Masters degree and Ph.D. degree in Mechanical Engineering from the Massachusetts Institute of Technology. He is currently an Assistant Professor at the Department of Mechanical Engineering in Chiang Mai University, Thailand teaching in energy related fields. He is originally from Ireland but enjoys the lifestyle in Northern Thailand. For the past several years he has been a consultant to the Energy Research and Development Institute Nakornping, Chiang Mai University. He has been heavily involved in carbon credit scheme for rural farmers, the conversion from liquefied petroleum gas to biomethane and drafting of a technical standard for a local biomethane gas grid. His research interests include biogas upgrading, biomethane grids, biomethane low pressure storage, meso-scale combustion, aerosol generation and pollution modeling.

All Authors have published several pier reviewed papers on biomethane and this book represents an agglomeration of this work.

Nomenclature

$\bar{P}_1$	Membrane permeability, (Barrer)
δ_m	Membrane thickness, (m)
$\dot{m}$	Mass flow rate, (kg/s)
$\dot{n}$	Molar flow rate, (mol/s)
$\dot{P}$	Electrical power, (kW)
$\dot{Q}$	Cooling load, (kW) or (kJ/h)
$\dot{V}$	Volume flow rate, (m^3/s) or (m^3/h)
μ	Dynamic viscosity, (kg/m.s)
ρ	Density, (kg/m^3)
$\vec{F}$	Force, (N)
$\vec{v}$	Fluid velocity, (m/s)
a_p	Interfacial area, (m^2/m^3)
c	Molar concentration, (mol/m^3)
C_{water}	Specific heat capacity of water, (J/kgK)
D	Diffusion coefficient, (m^2/s)
d_p	Particle diameter, (m)
E	Energy, (J)
E_f	Pipeline longitudinal joint factor, (–)
F	Pipeline design factor, (–)
f_D	Darcy friction factor, (–)
F_p	Packing factor, (m^{-1})
GWP	Global Warming Potential, (–)
H_G	Henry's constant, (kPa)
H_v	Volumetric heat rate, (J/m^3)
J_i	Flux, $m^3_{STP}/m^2.s$
k	Mass transfer coefficient, (1/s)
M	Molecular weight, kg/kmol
N	Mass transfer flux, (mol/s)
P	Pressure, (Pa)
P_i	Partial pressure of the *i*th constituent, (Pa)

R	Gas constant, (J/mol.K)
Re	Reynolds number, (–)
S	Pipeline yield strength, (kPa)
S_g	Gas-specific gravity, (–)
S_h	Pipe hoop stress, (kPa)
S_i	Sorption coefficient, ($m^3_{STP}/m^3.Pa$)
S_L	Laminar burning velocity, (m/s)
Sc	Schmidt number, (–)
SV	Space velocity, (h^{-1})
T	Temperature, (K)
t	Pipeline wall thickness, (mm)
u	Superficial velocity, (m/s)
W	Water concentration in desiccant, (mol/m^3)
WI	Wobbe Index, (–)

List of Figures

Fig. 1.1 Global installed electricity from biogas (Reprinted with permission from [14]) . . . 2
Fig. 1.2 Gross global energy consumption by fuel in 2014 (Reprinted with permission from [4]) . . . 2
Fig. 1.3 Distribution of electrical grid connected biogas production plants in Thailand [3] . . . 6
Fig. 1.4 Biogas production trend to 2022 in different areas of the world (Reprinted with permission from [11]) . . . 8
Fig. 1.5 Biomass use in primary energy supply in 2013 (Reprinted with permission from [5]) . . . 8
Fig. 1.6 Production of biogas and biomethane in Sweden (Reprinted with permission from [10]) . . . 9
Fig. 1.7 Biogas plants in Germany [18] . . . 10
Fig. 1.8 Chiang Mai University garbage/biomass management facility . . . 12
Fig. 1.9 30 tons per day municipal solid waste distribution through system . . . 13
Fig. 1.10 Select photos of the Chiang Mai University garbage management facility . . . 14
Fig. 2.1 The Thiobacillus thioparus shape . . . 27
Fig. 2.2 Thiobacillus thioparus after gram dye (100 times magnification) . . . 27
Fig. 2.3 Biofilter schematic diagram . . . 29
Fig. 2.4 Biofilter installation at a livestock farm in Thailand . . . 29
Fig. 2.5 Particle separator . . . 30
Fig. 2.6 Air blower . . . 31
Fig. 2.7 Biofilter columns . . . 31
Fig. 2.8 Pall rings filled inside the biofilter . . . 32
Fig. 2.9 Strainer . . . 32
Fig. 2.10 Centrifugal nutrient pump . . . 33
Fig. 2.11 Control panel housing . . . 33

Fig. 2.12 Control panels 34
Fig. 2.13 Biogas blower 34
Fig. 2.14 Post treatment pond 35
Fig. 2.15 Generalized pressure drop correlation from [36] 37
Fig. 2.16 Single-stage biogas dehumidification system 40
Fig. 2.17 Chiller unit for biogas dehumidification 43
Fig. 3.1 Scrubber for carbon dioxide removal in a non-regenerative wash 48
Fig. 3.2 Scrubber for carbon dioxide removal in a regenerative wash . . . 48
Fig. 3.3 Water solubility of carbon dioxide at various temperatures and pressures from National Institute of Standards and Technology (NIST) [11] 52
Fig. 3.4 Biogas upgrading using water scrubbing (Adapted with permission from [12]) 53
Fig. 3.5 Recovering hydraulic energy in biogas upgrading, adapted with permission from [13] 53
Fig. 3.6 Adsorption of CO_2 and CH_4 in carbon molecular sieve 3K at 298 K: **a** adsorption equilibrium; **b** uptake rate curves (Adapted with permission from [14], copyright 2005 American Chemical Society) 55
Fig. 3.7 Schematic of a pressure swing absorption and example of a PSA plant 56
Fig. 3.8 Chemical absorption process for biomethane production 57
Fig. 3.9 Physical solvent absorption process 60
Fig. 3.10 Carbon dioxide capture with a bubble column (Adapted with permission from [27]) 60
Fig. 3.11 Gas distribution through a membrane 61
Fig. 3.12 Membrane separation of carbon dioxide with an aqueous solution 62
Fig. 3.13 Typical gas separation membrane structure 62
Fig. 3.14 CO_2/CH_4 separation properties of various emerging polymer materials reported in the literature (Reprinted with permission from [29]) 63
Fig. 3.15 Range of biogas production costs in SE Asia and Europe for different biogas plants 66
Fig. 3.16 Specific capital costs for biomethane upgrading processes from two separate sources 68
Fig. 3.17 Total cost of upgrading to biomethane for five technologies 68
Fig. 3.18 Cost of biomethane production, compared to the price of natural gas 69
Fig. 3.19 Growth in biogas upgrading units distributed by technology employed, (Reprinted with permission from [28] (data from 2013 was not available)) 71
Fig. 3.20 Vapor compression refrigeration system 73

Fig. 3.21 Refrigeration chiller unit 73
Fig. 3.22 Solid desiccant properties, at atmospheric pressure and a 25 °C temperature. Adapted with permission from [33] 74
Fig. 3.23 The partial pressure of water vapor in equilibrium with the silica gel, absorption of water at various levels (Adapted with permission from [33]) 76
Fig. 3.24 Silica gel use in the gas dehumidification process (Adapted with permission from [33]) 77
Fig. 3.25 Molecular sieves 78
Fig. 3.26 Process diagram for moisture reduction in natural gas, with molecular sieve (Adapted with permission from [33]) 78
Fig. 3.27 General installation features for natural gas odorizers (with permission from King Tool Co., http://www.kingtoolcompany.com/wp-content/uploads/2012/11/Gas-Odorizers.pdf) 82
Fig. 3.28 Biomethane production plant 84
Fig. 3.29 Schematic diagram of a pressurized water biogas scrubbing system [36] 85
Fig. 3.30 Water flow rate and pressure versus biomethane purity 88
Fig. 3.31 Flash tank pressure versus final methane content 88
Fig. 3.32 Components of the biomethane water scrubbing system 91
Fig. 3.33 Components of the biomethane post treatment chilling system 91
Fig. 3.34 Desiccant cylinders for biomethane drying 93
Fig. 3.35 Plant civil engineering design 94
Fig. 3.36 Electrical control room for plant 94
Fig. 3.37 Gas analyzer and absorber liquid level sensor 95
Fig. 4.1 Summary of the GHG emission calculations including scenarios with the highest and lowest emissions for each waste type (Adapted with permission from [3]) 100
Fig. 4.2 Transportation fuel CO_2 Emissions per unit energy produced (Adapted from IPCC Guidelines for National Greenhouse Gas Inventories [2]) 102
Fig. 4.3 RBF; 3 ton per day, dedicated compressed biomethane (CBG) filling station produced from waste in Korat, Thailand 102
Fig. 4.4 Selected photos of CNG vehicles and filling stations 105
Fig. 4.5 Diagram of natural gas or biomethane injection system for vehicles 106
Fig. 4.6 Small-scale biomethane water scrubbing system 107
Fig. 4.7 CNG vehicle used in fuel comparison testing 108
Fig. 4.8 Horsepower and torque for three different fuels 109
Fig. 5.1 LPG total demand and use in domestic and industrial applications for Thailand, from [1] 112

Fig. 5.2 Biogas storage tank, regulator, and stove (Reprinted with permission from [11]) . . . 113
Fig. 5.3 Swirl and radial cooking stoves (Reprinted with permission from [11]) . . . 114
Fig. 5.4 Swirl burner head and inlet manifold (Reprinted with permission from [11]) . . . 114
Fig. 5.5 Cutaway view of fuel/air flow through the mixing chamber (Reprinted with permission from [11]) . . . 114
Fig. 5.6 Energy content in a 100 L tank of biomethane . . . 115
Fig. 5.7 Wobbe Index for biomethane with varying methane content (Reprinted with permission from [11]) . . . 116
Fig. 5.8 Average exit velocities through nozzle 1 (top) and nozzle 2 (bottom) (Reprinted with permission from [11]) . . . 120
Fig. 5.9 Radial and swirl burner optimal settings for biomethane, (Reprinted with permission from [11]) . . . 122
Fig. 5.10 Biomethane combustion at high flow settings for **a** swirl and **b** radial stoves (Reprinted with permission from [11]) . . . 123
Fig. 5.11 A High-temperature shuttle kiln . . . 125
Fig. 5.12 Rocket- and shower-type nozzle heads (Reprinted with permission from [20]) . . . 126
Fig. 5.13 Energy flow through nozzle versus the exit velocity for biomethane and LPG (Reprinted with permission from [20]) . . . 127
Fig. 5.14 Biomethane substitution results in a shuttle kiln (Reprinted with permission from [20]) . . . 128
Fig. 5.15 (1) LPG flame, 6.9 kPa, 0.9 mm, 2.9 kW: (2) Biomethane flame, 35 kPa, 1.4 mm, 3.1 kW (Reprinted with permission from [20]) . . . 129
Fig. 5.16 Temperature versus distance for biomethane and LPG flame (Reprinted with permission from [20]) . . . 129
Fig. 5.17 Emissions of CO and NO_x from LPG and biomethane (Reprinted with permission from [20]) . . . 130
Fig. 5.18 Price of LPG for industrial use in Thailand . . . 131
Fig. 5.19 Two 100 L tanks of CBG (Type 1) at 20 MPa (Reprinted with permission from [20]) . . . 132
Fig. 5.20 Production costs of CBG as a function of production capacity from [25] . . . 132
Fig. 5.21 Ceramic kiln with greenware and a volume of 0.1 m^3 (Reprinted with permission from [20]) . . . 133
Fig. 5.22 Temperature curve for firing greenware (Reprinted with permission from [20]) . . . 134
Fig. 5.23 Comparison of final products fired with LPG and biomethane (Reprinted with permission from [20]) . . . 134
Fig. 6.1 Area of a single class location unit . . . 146

Fig. 6.2 Construction of biogas grid pipeline 157
Fig. 6.3 Layout of pipeline for biogrid system 157
Fig. 6.4 Photos of different components used throughout the biomethane grid network 160
Fig. 6.5 Meters, shutoff valves, and drain valves throughout the local gas pipeline 161
Fig. 7.1 Biomethane adsorption relative mass: (kg biomethane gas/kg adsorbent) versus adsorbent pressure at 15 °C (Reprinted with permission from [24]) 171
Fig. 7.2 Biomethane adsorption: (grams of biomethane per tank liter) versus pressure at 15 °C (Reprinted with permission from [24]) 171
Fig. 7.3 Methane composition (CH_4) after desorption (% by Vol) versus number of cycles (Reprinted with permission from [24]) 172
Fig. 7.4 Schematic for liquid biomethane production 173
Fig. 7.5 ERDI's road map for biomethane research and development . . . 176

List of Tables

Table 1.1 General composition of biogas . . . 2
Table 1.2 Selected gas properties at atmospheric pressure and a temperature of 0 °C . . . 3
Table 1.3 Selected fuel properties . . . 4
Table 1.4 Natural gas standards for gaseous components . . . 5
Table 1.5 Emissions from certain transportation fuels . . . 5
Table 1.6 Plant electrical capacity for various regions of Thailand in June 2018 [3] . . . 7
Table 1.7 Biomethane production and infrastructure in Europe . . . 10
Table 1.8 Production of biomethane in European countries (Reprinted with permission from [1]) . . . 11
Table 1.9 Landfills biogas potential in the USA (Reprinted with permission from [8]) . . . 12
Table 1.10 Composition of biogas from CMU waste plant . . . 13
Table 2.1 Acceptable levels of hydrogen sulfide across selected applications . . . 18
Table 2.2 Hydrogen sulfide levels in biomethane for selected countries . . . 18
Table 2.3 Comparison of H_2S reduction methods . . . 18
Table 2.4 Biomethane maximum oxygen levels in selected countries . . . 19
Table 2.5 Biomethane moisture content in selected countries . . . 19
Table 2.6 Biomethane siloxane content in selected countries . . . 20
Table 2.7 Biogas pretreatment selection criteria . . . 21
Table 2.8 The bacteria strains which can remove hydrogen sulfide from biogas [9, 10] . . . 21
Table 2.9 Type of bacteria capable of removing hydrogen sulfide from various sources . . . 22
Table 2.10 Studies related to H_2S removal by biological process . . . 24
Table 2.11 More studies related to H_2S removal by biological process . . . 25

Table 2.12 More studies related to H_2S removal by biological process . . . 26
Table 2.13 H_2S removing performance by different bacteria . . . 28
Table 2.14 Physical properties of selected refrigerants . . . 41
Table 3.1 Biogas plant construction costs in SE Asia and Europe, from [30] . . . 65
Table 3.2 Biogas plant operation costs in SE Asia and Europe, from [30] . . . 65
Table 3.3 Average cost of upgrading to biomethane for different upgrading processes. (Reprinted with permission from [26]) . . . 67
Table 3.4 Compressed biomethane power requirements . . . 70
Table 3.5 Advantages and disadvantages of technologies to extract carbon dioxide . . . 71
Table 3.6 Composition of silica gel . . . 75
Table 3.7 Constants for Ergun's equation . . . 79
Table 3.8 Device details for Fig. 3.27 (King Tool Co.) . . . 83
Table 3.9 Biogas properties from swine farms . . . 85
Table 3.10 Biomethane properties from the absorber exit . . . 86
Table 3.11 Water properties in the water scrubber . . . 86
Table 3.12 Final specifications for the absorption tank . . . 89
Table 3.13 Final specifications for the flash drum . . . 90
Table 3.14 Final specifications for the desorption tank . . . 90
Table 3.15 Dehydrator specifications . . . 92
Table 3.16 Specifications for the gas analyzer . . . 95
Table 3.17 Specifications for the gas leak detector . . . 96
Table 4.1 Estimated number of CNG vehicles in Asia Pacific and top ten CNG countries worldwide (July 2018) . . . 100
Table 4.2 Biomethane production and infrastructure in Europe (EBA Biogas and Biomethane Report [5]) . . . 101
Table 4.3 Specifications on the nature and quality of biomethane for vehicles. Announced by Thailand's Department of Energy in 2018 . . . 103
Table 4.4 Requirements, limit values, and related test methods for natural gas and biomethane as automotive fuels . . . 104
Table 4.5 Fuel economic performance for gasoline, CNG, and biomethane . . . 109
Table 5.1 Comparison of laminar burning velocity for different fuels (Reprinted with permission from [11]) . . . 118
Table 5.2 Thermodynamic properties of both the biomethane and LPG . . . 118
Table 5.3 Optimal parameters for biomethane combustion . . . 123
Table 5.4 Stove energy output and efficiency at optimal conditions (Reprinted with permission from [11]) . . . 124

Table 5.5 Capital costs for installing LPG (Reprinted with permission from [20]) . . . 131
Table 5.6 Capital costs for installing a CBG system (Reprinted with permission from [20]). . . . 131
Table 5.7 Annual fuel savings for a CBG system (Reprinted with permission from [20]). . . . 135
Table 5.8 Economic analysis of changing over to CBG in a shuttle kiln (Reprinted with permission from [20]) . . . 135
Table 6.1 Applicable common requirements and test methods for biomethane at the point of entry into high-calorific gas and low-calorific natural gas networks. . . . 141
Table 6.2 Minimum methane content in biomethane for some European Countries [6] . . . 141
Table 6.3 Properties of biomethane for grid injection (high pressure, L-value) in the Netherlands [8] . . . 142
Table 6.4 Characteristics of natural gas for use in Singapore [9] . . . 142
Table 6.5 Advantages and disadvantages of using differing biomethane purity levels in local gas grids . . . 143
Table 6.6 Proposed list of components and their limits proposed in the local grid standard . . . 144
Table 6.7 Design factors for different class locations (Reprinted with permission from [2]). . . . 146
Table 6.8 Longitudinal joint factor (E_f) for steel pipe . . . 147
Table 6.9 Temperature derating factor (T) for steel pipe (Reprinted with permission from [2]). . . . 147
Table 6.10 Selected long-term hydrostatic strength (S) for certain thermoplastic materials. . . . 148
Table 6.11 Minimum wall thicknesses for reinforced thermosetting plastic pipe. . . . 148
Table 6.12 Minimum cover requirements for biomethane grid pipeline . . . 149
Table 6.13 Inspection of pipeline inspection frequency . . . 154
Table 6.14 Layout and diameter of biogrid pipelines for different outlets . . . 156
Table 6.15 PE 100 HDPE pipe dimensions . . . 158
Table 6.16 Layout and diameter of biogrid pipelines for different outlets . . . 159
Table 7.1 Substrate potential estimations in billion m^3/year for biomethane, (Reprinted with permission from [4]) . . . 164
Table 7.2 Properties of the absorbents used (Reprinted with permission from [24]) . . . 170
Table 7.3 Cost estimation of CBG v LBM in Thailand . . . 174
Table 7.4 R&D pathways for biogas production . . . 176
Table 7.5 R&D pathways for biomethane production . . . 177

Chapter 1
Introduction to Biomethane

1.1 Background

In 1630, Jan Baptist van Helmont discovered that organic material in decomposition produced flammable gases. Some years later (1776), Alessandro Volta discovered methane by collecting gas emerging from Lake Maggiore in Italy. In 1804, John Dalton established the chemical composition of methane. Louis Pasteur reported that biogas could be used for heating and lighting. The concept of anaerobic digestion was introduced around 1870 with the development of the septic tank system by Jean-Louis Mouras.

In the modern era, within the next few decades, bioenergy from biogas and biomethane has the potential to become a significant global renewable energy source as an economical attractive alternative to fossil fuels. The future potential of biomethane will be aided from its broad variety of applications, such as the production of heat, steam, electricity, hydrogen, and for use in transportation. Biogas can be produced in small and large scales which allows for versatile production all over the world. The electricity capacity from biogas plants in different regions of the world is given in Fig. 1.1.

Biogas production is predicted to increase to 40.2 million tons by 2030. The world bioenergy association estimated that renewable energy contributed approximately 18.6% of the total global energy consumption in 2014, in which bioenergy accounted for nearly 14% (Fig. 1.2).

1.2 Biogas Composition

Biogas consists of two main components, methane (CH_4) and carbon dioxide (CO_2), which is a nonflammable gas. Other components include hydrogen sulfide (H_2S), nitrogen (N_2) and oxygen (O_2). The general concentration of the gas is shown in Table 1.1.

S. Koonaphapdeelert et al., *Biomethane*, Green Energy and Technology,
https://doi.org/10.1007/978-981-13-8307-6_1

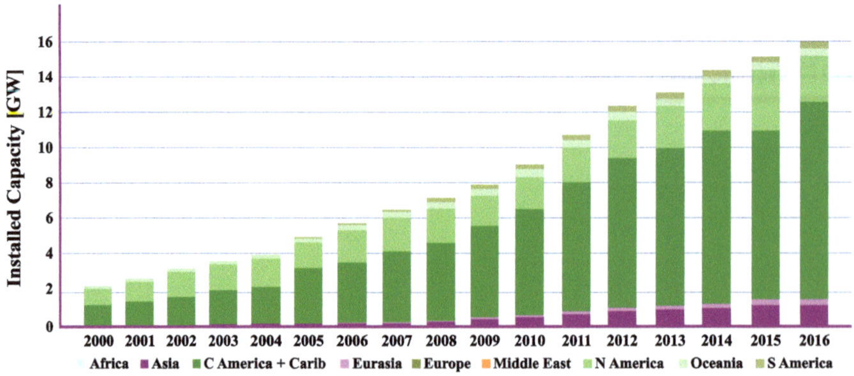

Fig. 1.1 Global installed electricity from biogas (Reprinted with permission from [14])

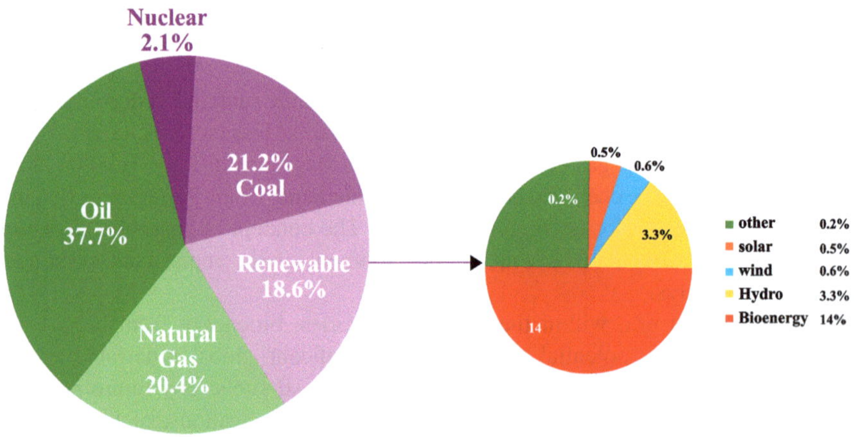

Fig. 1.2 Gross global energy consumption by fuel in 2014 (Reprinted with permission from [4])

Table 1.1 General composition of biogas

Biogas composition	Concentration levels
Methane (CH_4)	50–80% by Vol.
Carbon dioxide (CO_2)	20–50% by Vol.
Ammonia (NH_3)	0–300 ppm
Hydrogen sulfide (H_2S)	50–5000 ppm
Nitrogen (N_2)	1–4% by Vol.
Oxygen (O_2)	<1% by Vol.
Moisture (H_2O)	Saturated 2–5% by mass

Table 1.2 Selected gas properties at atmospheric pressure and a temperature of 0 °C

Biogas composition	(CH_4)	(CO_2)	(H_2)	(H_2S)	60% CH_4, 40% CO_2	65% CH_4, 34% CO_2, 1% Other
Heating value (MJ/m^3)	35.64	–	10.8	22.68	21.6	24.48
Ignition ratio (% air)	5–15	–	4–80	4–45	6–12	7.7–23
Ignition temperature (0 °C)	650–750	–	585	–	650–750	650–750
Change of state pressure (MPa)	4.7	7.5	1.3	8.9	7.5–8.9	7.5–8.9
Transition temperature (0 °C)	–82.5	31	–240	100	–82.5	–82.5
Density (kg/m^3)	0.72	1.98	0.09	1.54	1.2	1.15
Heat capacity (kJ/m^3/°C)	1.6	1.6	1.3	1.4	1.6	1.6

Biogas's usefulness as a fuel comes primarily from its methane component. It exists in the gaseous state at normal temperature, 0 °C (273.15 K), and pressure, 1 atm (101.325 kPa). If biogas liquefaction is desired, a pressure of approximately 20 MPa and temperatures of −161 °C are needed. Table 1.2 shows the heat output for different components and compositions of biogas. It has the ability to replace fossil fuels in transportation or for electricity generation [17].

1.2.1 Biogas and Pollution Reduction

Biogas can be formed from the organic compounds commonly found in wastewater from industrial plants, such as tapioca factories, breweries, fruit processing plants, as well as wastewater from livestock farms. Without biogas production, this organic waste is usually discarded into a nearby water supply, especially in developing countries. This effluent competes with downstream organisms for oxygen. This oxygen demand can kill marine life downstream of the discharge area. A measure of water quality is a parameter called the chemical oxygen demand (COD) which measures the amount of oxygen in the water consumed by organic reactions. It is expressed in mass of oxygen consumed over water volume, milligrams per liter (mg/L). A COD test can be used to quantify the amount of organic matter in water. It is a useful test because it provides a metric to determine the effect organic waste will have on the waterway. If instead of directly discarding the waste, biogas is produced, this reduces the wastewater chemical oxygen demand (COD) by over 80% [13]. In other words, the waste output from biogas production causes substantially less damage to the natural environment than the raw waste. A rule of thumb is that every kilogram of COD removed from wastewater can produce 0.3–0.5 m^3 of biogas.

Table 1.3 Selected fuel properties

Energy source	Density @ STP (kg/m^3)	Fuel heating value (kJ/kg)	Ignition temperature (°C)	Air/fuel ratio (kg/kg)	Octane number
Methane	0.72	50,000	650	17.2	100
LPG	540	46,000	400	15.5	30
Propane	2.02	46,300	470	15.6	35
Butane	2.70	45,600	365	15.6	10
Petrol	750	43,000	220	14.8	–
Diesel	850	42,500	220	14.5	–
Natural gas	0.83	57,500	600	17.0	80
Biogas (60% CH_4)	1.2	18,000	650	10.2	130

In addition to reducing pollution, the use of biogas instead of fossil fuels brings sustainable benefits. In rough numbers, the energy output from 1 m^3 of biogas could replace 0.46 kg of LPG, 0.67 liters of gasoline, 0.55 liters of fuel oil, or 1.2–1.4 kWh of electric power [6, 16]. Selected fuel properties are shown in Table 1.3.

1.3 Biomethane

1.3.1 Biomethane Definition

Biomethane is a gas that results from a process that improves the quality of biogas by reducing its levels of carbon dioxide, hydrogen sulfide, moisture, and other gases. Biogas upgraded to biomethane has a higher percentage of pure methane. It is sometimes called Renewable Natural Gas (RNG) and if compressed and used as a vehicle fuel it is known as Compressed Biomethane Gas (CBG). This is comparable to Compressed Natural Gas (CNG) or Natural Gas for Vehicles (NGV) which is derived from natural gas. Natural gas is a nonrenewable fossil fuel composed primarily of methane (CH_4, 70–98%) and traces of other hydrocarbons such as ethane, propane, butane, etc. Presently, there is no exact definition or standard that defines biomethane composition. However, the composition of biomethane is primarily natural gas and there exists natural gas standards written by agencies such as The Society of Automotive Engineers J1616 (SAE J1616.), California Air Resources Board (CARB), New Zealand Standard (NZS), and the California Public Utilities Commission (CPUC) as shown in Table 1.4.

Table 1.4 Natural gas standards for gaseous components

Components	SAE J1616 (1994)	CARB (1992)	NZS 5442 (1999)	CPUC Rule 30 (2002)
CH_4	–	88% (at least)	–	–
C_2H_6	–	6% (max)	–	–
C_3+	–	3% (max)	–	–
C_4+	–	–	–	–
C_6+	–	0.2% (max)	–	–
N_2	–	–	–	–
CO_2	3% (max)	0.1% (max)	–	3% (max)
Inert gas ($CO_2 + N_2 + O_2$)	–	1.5–4.5%	–	4% (max)
Sulfur	8–30 ppm	16 ppm (max)	50 mg/m^3	0.75 g/100scf (max)
Methane number	–	–	–	–
Heating value	–	–	–	36.1–42.8 MJ/m^3 (max)
Specific gravity			0.8 (max)	
Wobbe index	48.5–52.9	–	46–52	±10%

Note All percentages expressed as a mole. Except as otherwise shown

Table 1.5 Emissions from certain transportation fuels

Components	Emission unit	Gasoline	Diesel	CNG (CBG)
CO	g/km	10.9	0.662	6.54
NOx	mg/km	559	507	504
SO_2	mg/km	3.5	21.6	3.5
VOC	mg/km	662	166	146
TPM	mg/km	15.8	68.3	3.2
PM_{10}	mg/km	15.5	68.2	3.1
$PM_{2.5}$	mg/km	7.1	55.6	1.4

1.3.2 Advantages of Biomethane

Biomethane offers many advantages:

1. It is a renewable energy source.
2. When burned, it emits less pollution compared with diesel or gasoline. The emissions from these fuels are compared in Table 1.5.
3. Biomethane can be produced from locally made biogas. This eliminates the need to transport the gas over long distances as in the case of natural gas. Figure 1.3 shows the distribution of biogas-producing plants in Thailand.
4. By-products from the production of biogas can be used or sold as natural fertilizer.

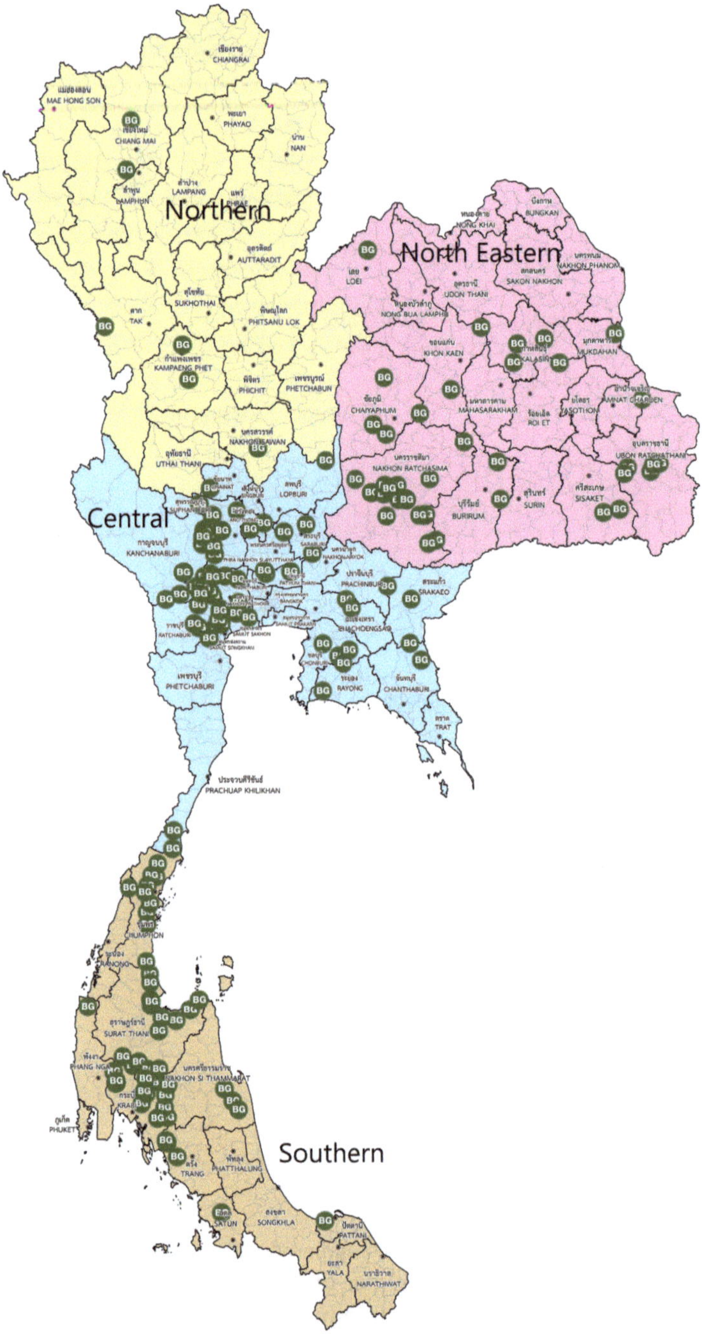

Fig. 1.3 Distribution of electrical grid connected biogas production plants in Thailand [3]

Table 1.6 Plant electrical-capacity for various regions of Thailand in June 2018 [3]

Region	Capacity (MW)
Northern	6
Northeastern	85
Central	87
Southern	121
Total	299

5. Organic waste from farms is sometimes disposed of in natural waterways causing pollution to marine life. Processing this waste into biomethane reduces this aquatic pollution.
6. An increased share of biomethane from within a country's own borders makes a nation's natural gas supply more reliable.
7. Biomethane is economically attractive, in terms of reducing the costs of importing fuel and increasing local employment in the production chain.
8. Rural areas especially profit from biomethane production because a considerable part of the revenue along the value chain is generated there.

The biogas plants shown in Fig. 1.3 distributed throughout the four regions of Thailand have an electrical capacity as shown in Table 1.6.

1.4 Biomethane Use Around the World

This section will give a short overview of biogas and biomethane production from three selected countries, Sweden, Germany, and the US. These countries have well-established biomethane industries. Their production potential and successful policies shall be briefly discussed.

According to the Intergovernmental Panel on Climate Change [9], greenhouse gas emissions must be reduced to less than half of the global emission levels of 1990. To take a step toward this goal, European energy production from biogas reached 6 million tons of oil equivalents (Mtoe) in 2007. The EU accounts for about 60% of the world's biogas production [12] and will likely keep its number one position, see Fig. 1.4. Globally, the biomethane market is still quite young. Some European countries have been active in biomethane production for years and have several plants in operation. Others have not implemented any biomethane projects so far.

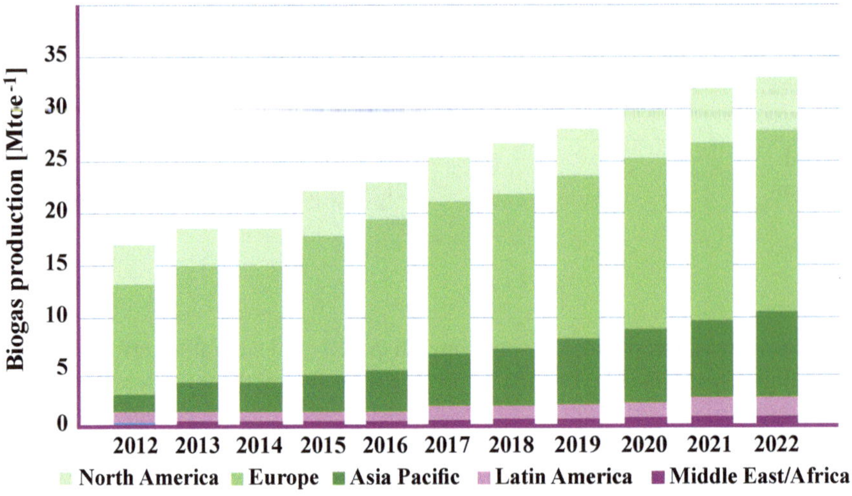

Fig. 1.4 Biogas production trend to 2022 in different areas of the world (Reprinted with permission from [11])

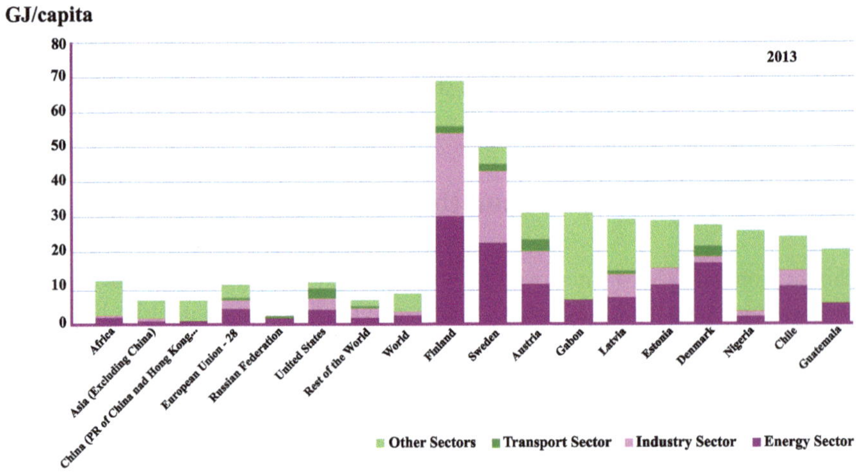

Fig. 1.5 Biomass use in primary energy supply in 2013 (Reprinted with permission from [5])

1.4.1 Sweden

In 2013, Sweden's share of renewable energy in its final energy use was the highest in the European Union at 52%. The Swedish use of biomass is exceptional among high-income countries. Only Finland uses more biomass for energy purposes on a per capita basis [5]. Sweden's use of biomass and waste accounted for 23% of their

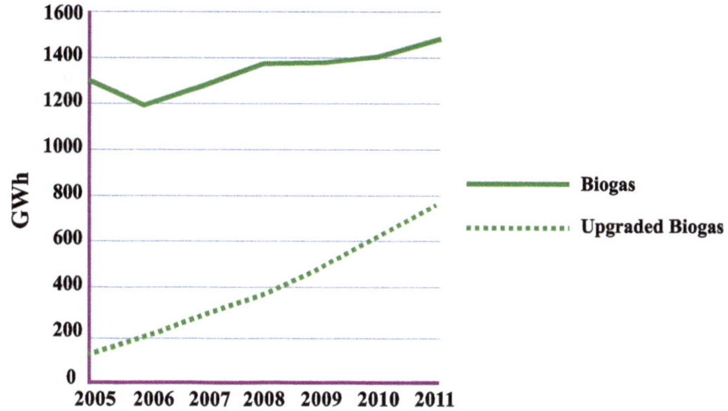

Fig. 1.6 Production of biogas and biomethane in Sweden (Reprinted with permission from [10])

national energy supply or 8% of all biomass and waste used in the European Union in 2013 (Fig. 1.5).

It is no surprise, therefore, that Sweden is one of the global leaders in biomethane production. Over half of all Swedish biogas is upgraded and that percentage is rising, see Fig. 1.6. Biomethane is predominately applied as a vehicle fuel in Sweden. It is processed and distributed through its natural gas pipelines and used for transportation purposes. As of 2016, Sweden had over 55,000 vehicles that could use biomethane along with a total of 226 refueling stations [7].

Biomethane could help reduce greenhouse gas emissions (GHG) from Swedish road transport by up to 25% [10]. The Swedish Government has set the target of attaining its vehicle fleet independent of fossil fuels by 2030 and a target of zero net greenhouse gas emissions by 2050 [15].

1.4.2 Germany

Germany produces biomethane for use as a transportation fuel and as a natural gas substitute in pipelines. Germany has become the largest biogas-producing country in the world, thanks to biogas plants on farms [19]. In 2008, there were around 4000 biogas production units on German farms, see Fig. 1.7. The combined electrical output from these plants is approximately 1.5 GW. The German government has set the year 2020 to produce 6 billion Nm^3 of biomethane as a natural gas substitute and 10 billion Nm^3 by 2030 [18].

The German policy is for GHG emissions in 2020 to be reduced by at least 40% from 1990 levels and by 2050 to be reduced by 80–90% from these levels [2]. In 2018, biogas and biomethane produced about 17% of electricity generation from renewable sources. Biomethane can be inserted into natural gas grids that crisscross Europe. This is known as feed-in. Germany has announced explicit targets for biomethane

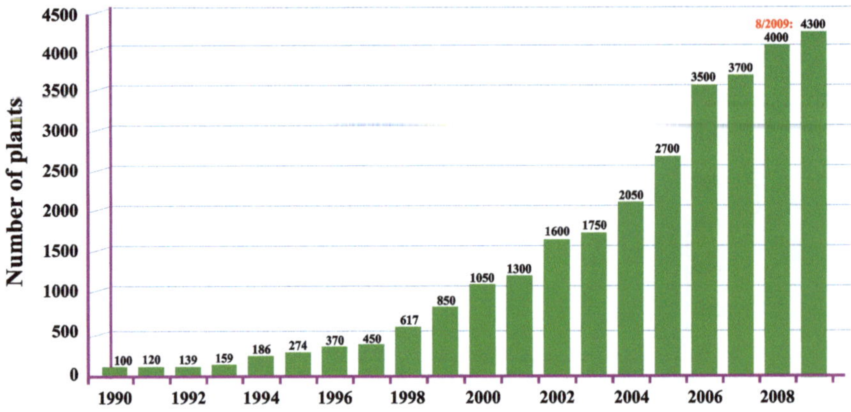

Fig. 1.7 Biogas plants in Germany [18]

feed-in to the German natural gas grid. The target mentioned is to replace 6% of national gas consumption with biomethane by the year 2020 and 10% by 2030. After feed-in, biomethane produced on one site can be used in geographically disperse applications. The situation is greatly helped by Germany having a well-developed gas pipeline network all over the country with a total length of 436,204 km (Tables 1.7 and 1.8).

Table 1.7 Biomethane production and infrastructure in Europe

Country	Biomethane plants	Raw biogas upgrading capacity (Nm^3/h)	Produced biomethane (GWh)	Plants feeding into national gas grid	Biomethane used in transport
Austria	14	5,160	70	11	n/a
Denmark	6	8,650	n/a	n/a	n/a
Finland	9	2,731	40	3	43%
France	8	2,610	41	6	n/a
Germany	178	204,082	9,140	165	3%
Hungary	2	625	4	1	n/a
Italy	5	500	n/a	n/a	n/a
Luxembourg	3	850	26	n/a	n/a
The Netherlands	21	16,720	683	n/a	n/a
Spain	1	4,000	n/a	n/a	n/a
Sweden	59	38,858	1,303	13	78%
Switzerland	24	6,310	166	22	33%
UK	37	18,957	700	34	n/a

Table 1.8 Production of biomethane in European countries (Reprinted with permission from [1])

	Number of feed-in plants (2012)	Total feed-in capacity (Nm^3/h)	Main market drivers	Dominating feedstock
Germany	110	66830	Government support schemes for energy crops and CHP	Energy crops Organic waste Manure
Sweden	7	2610	Vehicle fuel	Sewage sludge Organic waste
UK	2	160	Heat market	Organic waste
France	2	715	Government support schemes Vehicle fuel	Organic waste Sewage sludge
The Netherlands	17	7605	Green gas products Government support schemes	Organic waste Sewage sludge Landfill gas extraction
Switzerland	13	1664	Government support schemes Vehicle fuel	Sewage sludge Manure Organic waste

1.4.3 USA

The U.S. has over 2,000 sites producing biogas: 239 anaerobic digesters on farms, 636 landfill gas projects, and 1,241 wastewater treatment plants using an anaerobic digester, according to the U.S biogas council. There is much potential for biogas growth in the U.S. The U.S biogas council counts nearly 11,000 sites which could be commercially developed today: If fully realized, these biogas systems could produce enough energy to power 3.5 million American homes.

The energy potential of biogas in the United States was assessed at 18.5 billion m^3 of biogas/year, of which 7.3 billion m^3 comes from manure, 8.0 billion m^3 from landfill sites, and 3.2 billion m^3 from wastewater treatment plants that could generate about 41.2 TWh of electricity [14] (Table 1.9).

1.5 Case Study—Chiang Mai University Waste to Energy Plant

In 2017, Chiang Mai University implemented a garbage/biomass management facility located in the Mae Hia District of Chiang Mai. The facility consists of three major components: small-scale garbage sorting, anaerobic organic dry + wet fermentation, and a vehicle grade biomethane upgrading system as illustrated in Fig. 1.8. The main

Table 1.9 Landfills biogas potential in the USA (Reprinted with permission from [8])

Landfill output (SCFH)	Number of landfills	Total heat output (MW)
<6000	417	188.7
6000–21,000	320	711.8
21,000–42,000	130	861.3
42,000–72,000	171	1932.9
72,000–120,000	150	2847.0
120,000–300,000	188	7013.9
>300,000	98	12,193.3

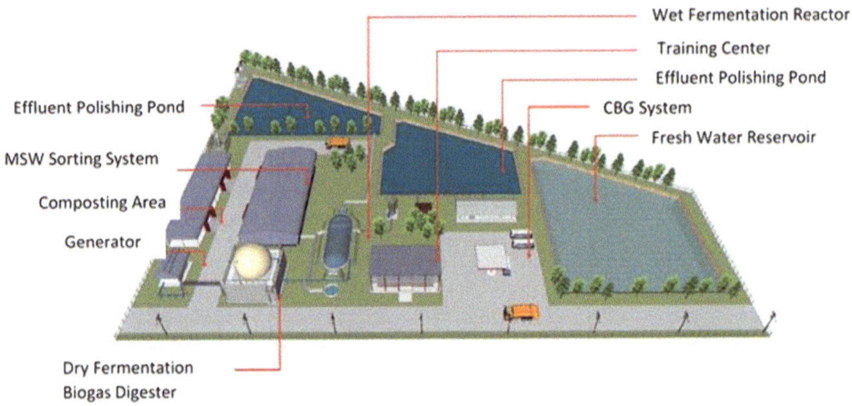

Fig. 1.8 Chiang Mai University garbage/biomass management facility

mission of this facility is to experiment and demonstrate suitable solutions to manage and utilize municipal waste in Thailand and Southeast Asia. The plant design and integration were completed by ERDI-CMU and construction and installation were completed and commissioned in January 2018.

The sorting equipment is supplied through Sanwa New Energy, a Chinese company. The system has two main components: a semi-manual picking station and an automatic ballistic separator. The throughput of the sorting equipment can achieve approximately 35% organic; 20% recyclables (including bottle, glass, and metal); 20% Refuse Derived Fuel (RDF) which is combustible; and 25% rejected inert matter which goes directly to landfills.

An approximate mass balance from a 30 ton per day municipal solid waste input is demonstrated in Fig. 1.9. In addition to the MSW, there are also small deliveries of agricultural waste (1–2 ton/day) and manure (3 ton/day). The sorting efficiency is expected to increase over time since source separation schemes will be implemented in Chiang Mai city in the near future. Separating the waste at source saves a lot of sorting time at downstream processes. The separated organic part is fed through an organic shredder before feeding into the main dry anaerobic digester. This digester

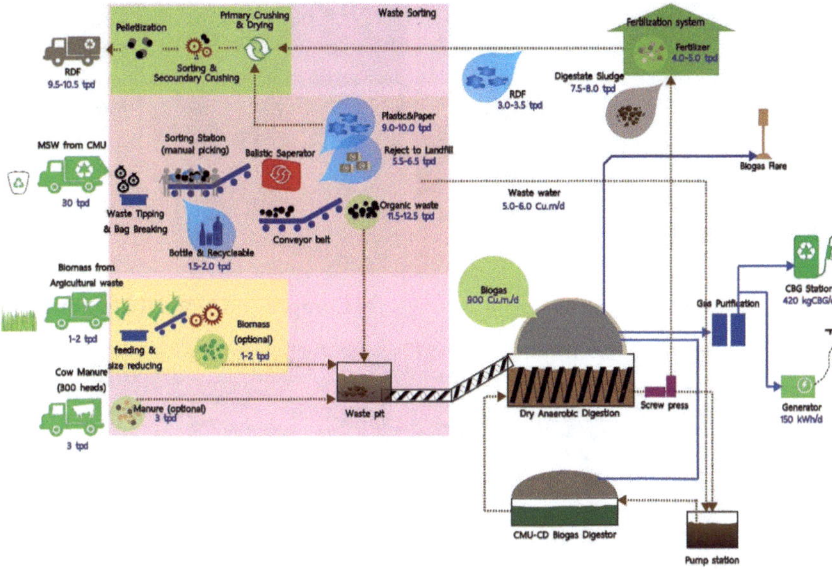

Fig. 1.9 30 tons per day municipal solid waste distribution through system

Table 1.10 Composition of biogas from CMU waste plant

Parameter	Quantity	Units
Temperature	25–35	°C
Pressure	2–3	kPa
CH_4	50–60	% Vol.
CO_2	35–45	% Vol.
H_2S	2,000	ppm
N_2	1–2	% Vol.
NH_3	<3,000	mg/m^3

has a total solid content above 15%. An additional plug-flow-type concrete lagoon wet digester was also installed to treat liquid effluent as well as additional wastewater from nearby facilities. The value of biogas yield from the dry fermentation process is expected to be in the range of 80–100 m^3 biogas/ton of sorted organic waste. So taking a 30 ton input per day with a 35% organic loading rate gives an approximate value for the biogas produced per day from Eq. 1.1 (Table 1.10).

$$30\ x\ 0.35\ x\ 100 = 1050\,\mathrm{m}^3\ biogas/day \tag{1.1}$$

Biogas produced from the facility is expected to be in the range of 800–1,200 m^3 per day from 20–30 ton MSW input operating at 10 h per day. Taking this untreated biogas to have a rough composition of 60% methane and 40% carbon dioxide means

(a) Plant digester

(b) Compressed biomethane storage

(c) Biomethane membrane upgrading station

Fig. 1.10 Select photos of the Chiang Mai University garbage management facility

a daily methane production rate of 480–720 m^3 per day. So the daily methane production in kg/day can be calculated from the ideal gas law (Fig. 1.10):

$$\frac{101325(480)}{\frac{8314}{16}(273)} \leq CH_4(kg/day) \leq \frac{101325(720)}{\frac{8314}{16}(273)} \tag{1.2}$$

$$340 \leq CH_4(kg/day) \leq 514 \tag{1.3}$$

This works out at an average of 430 kg of methane per day and 4.5 tons of organic fertilizer per day. The biogas is pretreated in a biological scrubber and evaporative dryer before separately fed to a 50 kW biogas genset and a 420 kg/day biomethane upgrading system. The upgrading system is a membrane separation system and a Chiang Mai University in-house design. The biomethane (87–90% CH_4 content) quality conforms to Thailand's Natural Gas for Vehicle (NGV) standard. The compressed biomethane (CBG) is utilized to fuel four public buses for intercampus transport and two garbage collection pickup trucks. The genset can be controlled by the campus power management center to function as backup and peak demand support. This plant is projected to achieve more than 5,000 tons of CO_2 equivalent emission reduction per annum. Nonetheless, this University's implementation is always open for site visits with training courses available to help convey the essential aspects of municipal solid waste management for a sustainable future.

References

1. Bowe S (2013) Market development and certification schemes for biomethane, chapter 19. Woodhead Publishing Series in Energy, pp 444–462. https://doi.org/10.1533/9780857097415.3.444
2. Daniel-Gromke J, Rensberg N, Denysenko V, Stinner W, Schmalfub T, Scheftelowitz M, Nelles M, Liebetrau J (2018) Current developments in production and utilization of biogas and biomethane in Germany. Chemie Ingenieur Technik 90:17–35. https://doi.org/10.1002/cite.201700077
3. Department of Alternative Energy. Home page (2018). http://weben.dede.go.th/webmax/
4. Kapoor R, Ghosh P, Kumar M, Vijay V (2019) Evaluation of biogas upgrading technologies and future perspectives: a review. Environ Sci Pollut Res 26:11631–11661
5. Karin E, Sven W (2016) The introduction and expansion of biomass use in Swedish district heating systems. Biomass Bioenergy 94:57–65
6. Kohl AL, Nielsen RB (1997) Gas purification, 5th edn. Gulf Professional Publishing, Houston
7. Lonnqvist T, Gronkvista S, Sandberg T (2017) Forest-derived methane in the Swedish transport sector: a closing window. Energy Policy 105:440–450
8. Murray BC, Galik CS, Tibor V (2014) Biogas in the United States: an assessment of market potential in a carbon constrained future. Technical report, Nicholas Institute for Environmental Policy Solutions, 2
9. Nakicenovic N et al (2000) A special report of working group III of the intergovernmental panel on climate change. Technical report, Intergovernmental Panel on Climate Change
10. Olsson L, Fallde M (2015) Waste(d) potential: a socio-technical analysis of biogas production and use in Sweden. J Clean Prod 98:107–115
11. Pike Research (2012) Methane recovery and utilization in landfills and anaerobic digesters: municipal solid waste, agricultural, industrial, and wastewater market report on analysis and forecasts. Technical report, Renewable biogas
12. Raboni M, Urbini G (2014) Production and use of biogas in Europe: a survey of current status and perspectives. Ambiente Agua 9(2):192–202. https://doi.org/10.4136/ambi-agua.1324
13. Satyawali Y, Balakrishnan M (2008) Wastewater treatment in molasses-based alcohol distilleries for cod and color removal: a review. J Environ Manag 86:481–497
14. Scarlat N, Dallemand J-F, Fahl F (2018) Biogas: developments and perspectives in Europe. Renew Energy 129:457–472
15. Tomas L, Alessandro S-P, Thomas S (2015) Biogas potential for sustainable transport - a Swedish regional case. J Clean Prod 108:1105–1114
16. von Mitzlaff K (1988) Engines for biogas. The Deutsches Zentrum für Entwicklungstechnologien, Sept
17. Walsh J, Ross C, Smith M, Harper S, Wilkins A (1988) Biogas utilization handbook. Georgia Tech Research Institute, Atlanta, The Environmental, Health and Safety Division
18. Weiland P (2009) Status of biogas upgrading in Germany. In: IEA task 37 workshop "Biogas upgrading", Johann Heinrich von Thunen-Institute, Federal Research Institute for Rural Areas, Forestry and Fisheries
19. Weiland P (2010) Biogas production: Current state and perspectives. Applied Microbiology and Biotechnology 85:849–860. https://doi.org/10.1007/s00253-009-2246-7

Chapter 2
Biogas Cleaning and Pretreatment

2.1 Contaminants

The following biogas components are commonly found at the digester exit and need to be removed before entering the most upgrading processes.

(A) Hydrogen Sulfide Hydrogen sulfide is one of the contaminants that needs to be lowered substantially. The exact requirements are dependent on the upgrading technology used. For example, in membrane and PSA upgrading systems, the required H2S level is <10 ppm. This contaminant is caused by bacteria decomposing inorganic sulfides under anaerobic conditions. Its elements are hydrogen and sulfur, H_2S. When exposed to water or steam it is converted into sulfuric acid (H_2SO_4) which causes metal corrosion. Reducing hydrogen sulfide before combustion is of benefit to the environment and helps extend the lifetime of the equipment [1] (Table 2.1).

Different countries allow different concentrations of hydrogen sulfide in their biomethane, as shown in Table 2.2. The European standard is a maximum value for both hydrogen sulfide and carbonyl sulfide (COS) [2, 3].

Authur and Anna [4] discussed methods to reduce hydrogen sulfide:

- Aeration (air/oxygen dosing method),
- The addition of iron chloride (Iron Chloride Dosing Method),
- The use of powder metal (iron sponge method),
- The use of lump iron (iron oxide pellets method),
- The use of carbon (activated carbon method),
- Absorption by water (water scrubbing method),
- Absorption by sodium hydroxide (NaOH scrubbing method),
- The use of microorganisms (biological removal method), and
- Aeration and recovery (air stripping and recovery method).

Each of these technologies has its own unique advantages and disadvantages, from different costs to efficiencies, see Table 2.3.

S. Koonaphapdeelert et al., *Biomethane*, Green Energy and Technology,
https://doi.org/10.1007/978-981-13-8307-6_2

Table 2.1 Acceptable levels of hydrogen sulfide across selected applications

Equipment	Max. H_2S concentration (ppm)
Fuel cell	1
Grid injection	5
Vehicle fuel	25
Electrical generator (CHP)	200
Boiler	1000 (usually raw biogas)

Table 2.2 Hydrogen sulfide levels in biomethane for selected countries

Country	Max. H_2S concentration (mg/m^3)
Europe (grid injection) ($H_2S + COS$)	5
Europe (vehicles) ($H_2S + COS$)	5
Thailand	23

Table 2.3 Comparison of H_2S reduction methods

Method	Removal efficiency	Capital cost	Operating cost	Complexity
Biological removal method	Moderate	Moderate	Low	Moderate
Iron chloride dosing	Moderate	Low	Moderate	Low
Water scrubbing	High	High	Moderate	High
Activated carbon	High	High	Moderate	Moderate
Iron sponge method	High	Moderate	Moderate	Moderate
NaOH scrubbing method	High	Moderate	High	Moderate

(B) Oxygen Oxygen is found in small quantities in raw biogas through anaerobic digestion. In the biological removal method, if air is used to remove hydrogen sulfide this also increases the amount of oxygen in the biogas. Too much oxygen is an explosion risk. If used in an IC engine, the oxygen can cause premature ignition or "knocking". Some of the upgrading techniques, such as PSA and membrane (see Sect. 3.1) have the ability to remove oxygen thereby negating the need for an additional process. If a specific oxygen removal process is needed then choices include adsorption by a carbon molecular sieve and catalytic oxidation using platinum and palladium catalysts and chemisorption using a copper surface [5]. After upgrading, the final biomethane must be compatible with government regulations. Different countries allow different concentrations of oxygen in their biomethane, as shown in Table 2.4 [2, 3].

(C) Humidity Raw biogas typically contains saturated water vapor. This dilutes the gas quality reducing its heating value. The humidity can be a problem if it forms ice, clogging pipes and valves or steam which can cause a pressure rise. Condensed water

Table 2.4 Biomethane maximum oxygen levels in selected countries

Country	Max. O_2 concentration (% by Vol.)
Europe (grid injection)	0.001–1
Europe (vehicles)	1
United States	3
Thailand	1

Table 2.5 Biomethane moisture content in selected countries

Country	Biomethane moisture conditions
Europe (grid injection)	Dew point −8 °C at 7,000 kPa
Europe (vehicles)	Dew point −10 to −30 °C at 20,000 kPa
Thailand	Dew point 4.4 °C at 20,000 kPa
United States	No condensation

can cause pipe erosion or react with hydrogen sulfide to form sulfuric acid. Moisture can penetrate the lumen of a membrane reducing its effectiveness quickly over time. Water scrubbing is a biogas upgrading technology that is unaffected by the moisture content in the biogas; however, the moisture must be removed post-operation. The most common method for removing moisture is refrigeration. Cooling the biogas below its dew point temperature causes the moisture to condense. Normally the cooling is followed by liquid removal in a knock-out vessel before reheating the biogas to ensure no condensation inside the upgrading equipment. Different countries allow different concentrations of moisture in the final product, as shown in Table 2.5 [2, 3].

(D) Ammonia Raw biogas from substrates rich in nitrogen can contain up to 200 ppm of ammonia. When ammonia combusts it causes nitrogen to form nitrogen oxides, a corrosive and environmentally unfriendly gas. Ammonia must be limited to 10 ppm before use in membrane upgrading processes as it can degrade the membrane. Ammonia can be removed by condensation drying since it can be dissolved. If water absorption is used, both moisture and ammonia are removed. Molecular sieves can lower ammonia to below 1 ppm in the PSA process. In cases with high ammonia concentrations, a gas wash dryer can be used Hoyer et al. [5]. Europe has adopted a final limit of 10 mg/Nm^3 for ammonia before biomethane can be injected into a gas transmission pipeline [6].

(E) Particulates Particles of dust along with oil mist from the biogas blower may also be found. These contaminants should be filtered with a 1–5-micron paper or fabric filter. For membrane and PSA separation, the total particulate and aerosol content should be kept below 0.01 mg/m^3 [7]. A course filter is needed immediately after the digester with a fine filter ($<1\ \mu$m) being placed just prior to the membrane

Table 2.6 Biomethane siloxane content in selected countries

Country	Biomethane siloxane level (mgSi/m^3)
Europe (grid injection)	0.3–1
Europe (vehicles)	0.3
Thailand	0.3

or PSA. This is why oil-free compressors are recommended for use in biomethane plants. Water scrubbing is more tolerant and can operate at higher particle levels.

(F) Siloxanes Siloxane compounds can be found in cosmetics, deodorant, soap, and food. Siloxanes in biogas mostly come from landfill gas and wastewater treatment plants. Siloxanes are rare in biogas produced from agricultural sources as animals do not use cosmetics. Biogas from landfill can have siloxane concentrations up to 50 mg/Nm^3. They are corrosive, reducing engine lifetimes which have a siloxane limit of 15 mg/m^3. They can be removed by activated carbon absorption, through organic solvents or hot sulfuric acid although these systems are relatively expensive. Cooling the gas can remove siloxanes and water simultaneously. However, this is not efficient as the system would need to operate at a temperature as low as −30 °C (Table 2.6).

(G) Nitrogen Nitrogen reduces the overall heat of combustion. Biogas from landfill waste contains a high proportion of nitrogen although it is rarely found in biogas from agricultural waste. Nitrogen removal is difficult since nitrogen is an inert gas. Pressure Swing Adsorption (PSA) can remove nitrogen simultaneously with carbon dioxide. Cryogenic cooling can be used to liquefy and remove the nitrogen but it is very expensive [8].

(H) Others Halogenated hydrocarbons are high molecular weight hydrocarbons usually found in biogas from landfill waste. They are not common in biogas from wastewater treatment plants or organic waste. They corrode and can form dioxins and furans. Activated carbon processing can remove halogenated hydrocarbons.

Volatile organic compounds (VOCs) similarly are found primarily in landfill gas at levels up to 700 mg/m^3. Energy crops and manure produce VOCs in lower quantities, 10–30 mg/m^3. VOCs can cause corrosion and react with metals. They can be removed partially by water scrubbing, membrane, and PSA technologies. If more thorough removal is necessary then activated carbon processing should be used.

2.2 Pretreatment Selection Criteria

The requirement for biogas pretreatment depends on an upgrading technique used. As can be seen in Table 2.7, most upgrading processes are able to remove only carbon dioxide from biogas. Pressure swing adsorption has an ability to remove oxygen and hydrogen sulfide at low concentration. Some contaminants are partially removed,

Table 2.7 Biogas pretreatment selection criteria

Upgrading techniques	Contaminants		
	Removed	Partially removed	Not removed
Water scrubbing	CO_2	VOC, H_2S, NH_3	O_2, N_2, H_2, H_2O
Amine scrubber	CO_2	VOC, H_2S, NH_3	O_2, N_2, H_2, H_2O
Membrane	CO_2	O_2, H_2S, H_2O, H_2	VOC, NH_3, N_2
PSA	CO_2, O_2, H_2S (low conc.)	N_2, H_2O	H_2, VOC, NH_3
Organic physical scrubber	CO_2	VOC, H_2S, NH_3	O_2, N_2, H_2, H_2O
Cryogenics	CO_2	VOC, H_2S, NH_3	O_2, N_2, H_2, H_2O

indicating a need for biogas pretreatment. Contaminants listed as "not removed" by a particular upgrading process are critical. There must be a method, either raw biogas pretreatment or biomethane post treatment, to control their concentrations in the final product.

2.3 H_2S Bioscrubber

This section details the removal of hydrogen sulfide with biological processes which is the method of choice since it has advantages over chemical processes when taking long-term energy consumption, operational costs, and environmental friendliness into account. Bacterial consortium which consume H_2S used in this process produce a variety of enzymes that accelerate their metabolic mechanisms. They also possess the ability to adapt themselves to survive in a changing environment. The bacteria usually comprise many species of heterotrophs and autotrophs as shown in Table 2.8.

Bacteria which have the ability to remove hydrogen sulfide are usually identified as sulfur-oxidizing bacteria. As can be seen in Table 2.8, they require different carbon sources and energy sources but always need a substrate which contains sulfur as the

Table 2.8 The bacteria strains which can remove hydrogen sulfide from biogas [9, 10]

Term	Energy source	Electron donor	Carbon source	Organism
Chemolithoautotroph	Chemical	H_2S	CO_2	Paracoccus, starkeya, beggiatoa, thiobacillus, thiomagarita, thioploca, thiomicrospira, sulfolobus, acidianus
Chemolithoheterotroph (mixotrophy)	Chemical	H_2S	Org. *C*	Beggiatoa, thiothrix, leucothrix
Photolithoautotroph	Light	H_2S	CO_2	Chlorobium, chromatium, allocromatium, thiocapsa, halochromatium, rubrivivax, rhodobacter, chloroflexus

Table 2.9 Type of bacteria capable of removing hydrogen sulfide from various sources

Source	Organism	Medium	pH	Reference
Activated sludge	Thiomonas sp.	Thiosulfate medium (TS medium)	7.0	[12]
Sulfide-rich wastewater	Alcaligenes sp.	Thiosulfate mineral salts medium (thiosulfate MSM)	6.0	[13]
Peat biofilter	Thiobacillus intermedius	Modified Waksman's medium	3.0	[14]
Soap lake (Washington State)	Thioalkalimicrobium microaerophilum	Mineral medium	10.0	[15]
Rhizospheric soil	Paracoccus bengalensis sp. nov.	Basal mineral salts solution (MS)	7.0	[16]

main component. Different species might require different sulfur oxidation numbers. The substrate selection plays a very important role in promoting the growth of sulfur-oxidizing bacteria [11]. The hydrogen sulfide is usually oxidized into elemental sulfur or insoluble sulfuric compounds.

Sulfur-oxidizing bacteria are obtained from many sources. Species of bacteria, of interest, can be separated from these sources by using a selective medium which only the sulfur-oxidizing bacteria can grow on. Some important bacteria, sources, and media are shown in Table 2.9.

2.4 Biofilter Reactor

A biofilter is a type of biological reactor for hydrogen sulfide removal. The main principle is to drive raw biogas through a packed column bed, either upward or downward, while keeping suitable temperature, moisture, and pH to promote the growth of sulfur-oxidizing bacteria [17]. The packing media used in the reactor are usually porous and coarse, allowing bacteria to attach. Suitable media properties include [18]:

1. Good moisture holding capacity,
2. Highly porous with high surface area per volume,
3. High compressive strength,
4. Low pressure drop,
5. Resistance to biodegradability,
6. Lightweight,
7. Relatively low cost,
8. Good absorption capacity, and
9. Resistance to acidity.

Natural materials such as peat, fiber, compost, and tree bark can be used as the media; however, their lifetimes are relatively short. On the other hand, synthetic materials, e.g., ceramics, activated carbons, and polystyrene packings, can be used as well although their costs are higher [19].

There are several factors affecting the H_2S removal performance including, temperature, moisture, biogas flow rate, concentration, and space velocity. The biogas flow rate and H_2S concentration are considered to be most influential factors in the design of a fixed bed-volume reactor. Duan et al. [20] studied the removal of hydrogen sulfide using biological process and compared the efficiency of hydrogen sulfide removal using different media, namely Biological Activated Carbon (BAC) and Virgin Activated Carbon (VAC) by using an upflow biofilter. The carbon bed had a height of 20 cm (medium volume = 0.2 L, specific surface area of carbon = $807\,m^2/g$). The gas retention time was in the range of 2–21 s, leading to a Space Velocity (SV) of $170–1200\,h^{-1}$. The study found that BAC had the ability to eliminate hydrogen sulfide gas better than the VAC. The efficiency of H_2S removal was 94–99%.

Where the space velocity is defined by

$$SV = \frac{Q}{V} \tag{2.1}$$

where

- SV = Space velocity (h^{-1})
- Q = Biogas flow rate (Nm^3/h)
- V = Media bed volume (m^3)

Lee et al. [21] conducted a downflow biofilter experiment, by comparing the effectiveness of three strains of *Thiobacillus thiooxidans,* AZ11, MET, and TAS. They adhered to the surface of the porous ceramic medium. The results showed that AZ11 had the highest H_2S removal efficiency compared to the others. Therefore, it was chosen to run in a long-term experiment in which the SV equaled to $200\,h^{-1}$, the inlet concentration equaled to 200–2200 ppmv and the inlet loading rate was 47–$670\,gs/m^3/h$. The system had a removal efficiency of 99.9%. However, when the space velocity exceeded 500 and $600\,h^{-1}$, the removal efficiency was lowered to 98% and 94%, respectively.

Ma et al. [22] compared the efficiency of 4 types of media: peat moss, wood chip, ceramic, and Granular Activated Carbon (GAC) for upflow biofilter tanks. The columns were made of Pyrex glass, using *Thiobacillus* denitrificans for H_2S removal and were sprayed with distilled water to maintain moisture inside the system. At an inlet concentration of 100 mg/l, it was found that GAC had the highest removal efficiency of 98%.

Aroca et al. [23] conducted a comparative study of H_2S removal in trickling filters by growing *Thiobacillus thioparus* and *Acidithiobacillus thiooxidans* on polyethylene rings. The study found that when the concentration of H_2S was higher, the gas flow rate in the system was lower. It was found that *A. thiooxidans* could eliminate 100% of H_2S while *T. thioparus* could eliminate 47% of H_2S.

Kim et al. [24] also studied the removal of H_2S by using bead in Pall rings which were made from sodium alginate and PVC. The experiment was conducted for 90 days by increasing the loading rate from 0.1 to 13 $gH_2S/m^3.h$. The results showed that the removal efficiency increased to 99% within the first 2–3 days. A shock load condition occurred, when the H_2S loading rate became too high, the efficiency of the system reduced to 78%. The system efficiency was restored by lowering the loading rate.

Rattanapan et al. [25] studied the removal of hydrogen sulfide by using Sulfide-Oxidizing Bacteria (SOB) from the wastewater of latex production plants. The bacteria were grown on a Granular Activated Carbon (GAC) media. It was found that the SOB biofilter could remove 98.7% of H_2S at an inlet concentration of 200 ppmv. The removal efficiency tended to increase slightly when the concentration increased from 200 to 4,000 ppmv. The overall results of these and other studies are summarized in Tables 2.10, 2.11, 2.12.

Table 2.10 Studies related to H2S removal by biological process

Reference	[19]	[18]	[20]	[25]
Bacteria	Mixed microbial culture	Nature	Sulfide oxidizing	Sulfide oxidizing
Packing material	Inorganic	Pig manure and sawdust	Activated carbon	Granular Activated Carbon (GAC)
Influent H2S concentration (ppmv)	2,000	–	5–100	200
H2S removal efficiency (%)	–	>90%	94–99%	>98%
H2S removal capacity ($g/m^3.h$)	–	10–45	110–181	125
Gas flow rate (m^3/h)	–	–	0.034–0.24	0.035
Space velocity (h^{-1})	–	–	170–200	52.24
Retention time (s)	–	13.5–27	2–21	60 days
pH (the growth of bacteria)	4.7	6.8–8.4	1.0–7.0	2.10–8.35
Temperature (°C)	–	20–22	25	27–32
Surface area of packing material (m^2)	–	–	807 m^2/g	–
Reactor type	Downflow	Downflow	Upflow	Downflow
Reactor dimension	$d_i = 5$ cm $H = 50$ cm	$d_i = 10$ cm $H = 1$ m	$d_i = 3.6$ cm $H = 30$ cm	$d_i = 5.5$ cm $H = 60$ cm
Moisture	70–80	–	–	–
Pressure drop	–	15–460 $Pa.m^{-1}$	–	–
Final product	–	Sulfate, thiosulfate	Sulfate, TS	Sulfur, sulfate

Table 2.11 More studies related to H2S removal by biological process

Reference	[26]	[27]	[20]	[23]	
Bacteria	*Thiobacillus thioparus CH11*	*Thiobacillus thioparus*	*Acidithiobacillus thiooxidans*	*Acidithiobacillus thiooxidans*	*Thiobacillus thioparus*
Packing material	Ca-alginate bead	Peat	Pellet activated carbon	Polyethylene ring	
Influent H2S concentration (ppmv)	5–100	355	<87	50	4600 and 98.2
H2S removal efficiency (%)	>98	100	>94	100	100
H2S removal capacity (g/m^3.h)	25	55	181	14	370
Gas flow rate (m^3/h)	0.018–0.185	0.03		0.07	0.03
Space velocity (h^{-1})	26–265	7.89	0.05–0.33	–	–
Retention time (s)	28	–	<2	26	120 and 45
pH (the growth of bacteria)	6.0–8.0	5.5–6.0	1.0–7.0	5.5–7	
Temperature (°C)	28 ± 2	30	25	30	
Surface area of packing material (m^2)	0.28	–	807	340	
Reactor type	Upflow	Downflow	Upflow	Downflow	
Reactor dimension	$D_i = 0.055$ m $H = 0.60$ m	$D_i = 0.055$ m $H = 0.60$ m	$D_i = 0.036$ m $H = 0.30$ m	$D_i = 0.055$ m, $H = 0.60$ m	
Moisture	95–100%	92%	42.40% (100 ml/day)	–	–
Final product	Sulfur, sulfide, sulfate, and thiosulfate	–	Sulfate and biomass	–	–

2.4.1 Selection of H_2S Removing Bacteria

There are two types of bacteria used in a biofilter, which are pure culture and mixed culture. A biofilter is sometimes started up by inoculating a pure culture on media; however, as the operation continues, other types of bacteria can also grow and cohabit as a mixed culture.

For pure culture bacteria, *Thiobacillus thioparus* is a sulfur-oxidizing bacteria group that has the ability to oxidize sulfur compounds (hydrogen sulfide, sulfide, sulfur, and thiosulfate), and this type of bacteria from the group chemoautotroph can synthesize food itself without energy from light. It grows well in a wide pH range of 5–9 [30]. The appropriate temperature range for its growth is 20–37 °C [26] which is common in tropical countries. This bacteria is available for purchase in many countries. For example, *Thiobacillus thioparus* (JCM 3859) which is small, opaque

Table 2.12 More studies related to H2S removal by biological process

Reference	[28]	[18]	[24]
Bacteria	*Thiobacillus thiooxidans KS1*	Mixed culture	Mixed culture
Packing material	Cylindrical ceramic	Pig manure and sawdust	The bead in pall ring (Na-alginate + PVC)
Influent H2S concentration (ppmv)	2.00 μmol/l and 1.34 μmol/l	–	10–130 ppmv
H2S removal efficiency (%)	Depends on packing material size	>90	<90
Gas flow rate (m^3/h)	180 m/h and 720 m/h (Volume flow velocity)		–
Space velocity	–	100	200
Retention time (s)	16 and 10	27	13.5
pH (the growth of bacteria)	–	6.8–8.4	–
Temperature (°C)	–	20–22	30
Reactor	Upflow	Downflow	Downflow
Reactor dimension	$D_i = 0.15\,m\ H = 0.80\,m$ and $D_i = 0.25\,m$, $H = 1.0\,m$	$D_i = 0.10\,m,\ H = 0.80\,m$	$D_i = 0.14\,m,\ H = 1.6\,m$
Moisture	0.8 L/min	–	0.5 L/min
Pressure drop	–	15–460 $Pa.m^{-1}$	–
Final product	H_2SO_4	Sulfur, sulfate, and thiosulfate	Sulfur, sulfide, sulfate, and thiosulfate
Reference	[29]	[21]	[22]
Bacteria	*Thiobacillus thiooxidans AZ11*	*Acidithiobacillus thiooxidans AZ11*	*Thiobacillus denitrificans*
Influent H2S concentration	200 mg/l	200–2000 ppmvv	100 mg/l
H2S removal efficiency (%)	99.99%	99.99%	98%
H2S removal capacity (Vm)	$428\,g - s/m^3h$	$47 - 670\,g - s/m^3h$	–
Space velocity	300/h	200/h	–
Retention time	–	9 s	30–100 s
pH	–	1.5	6.8–7.4
Temperature (°C)	30	–	30
Media	Porous lava (from Korea, types A and C)	Porous ceramic	Granular activated carbon
Surface area of media	1.41(Å) and 3.47 (C) m^2/g	–	1250 m^2/g
Reactor	Downflow	Downflow	Upflow
Pressure drop	–	–	11.9–45.2 mmH_2O/m
Moisture content	–	80–90%	–
Water used	200 ml/day	50 ml/day	50% Vol.
Reactor dimension	$D_i = 0.046\,m,\ H = 0.3\,m$	$D_i = 0.046\,m,\ H = 0.3\,m$	$D_i = 0.08\,m,\ H = 0.30\,m$

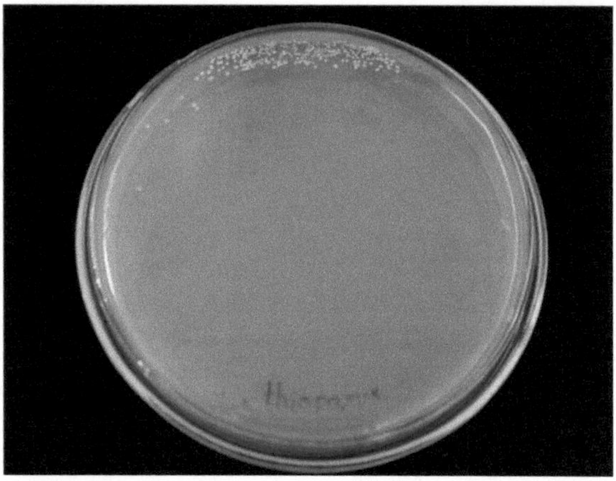

Fig. 2.1 The Thiobacillus thioparus shape

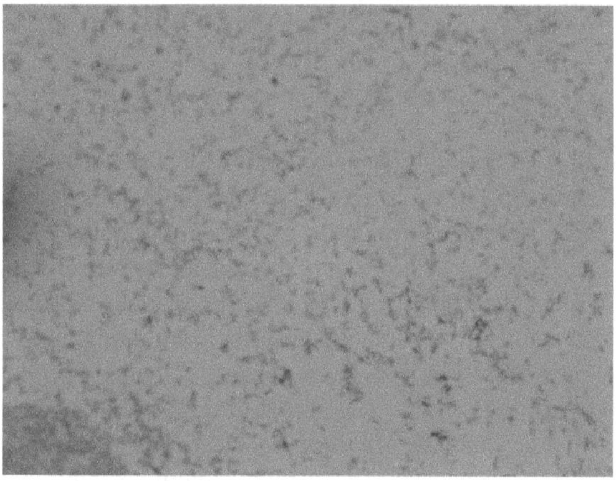

Fig. 2.2 Thiobacillus thioparus after gram dye (100 times magnification)

white, and cylindrical, as shown in Fig. 2.1, can be ordered from the Japan Collection of Microorganisms. It is a gram-negative as shown in Fig. 2.2.

H_2S bacteria can be collected from the biofilter packing material and grown in liquid Thiobacillus Enrichment Media (TS(N)) which can selectively separate sulfur-oxidizing bacteria from others. In TS(N) media, there are only thiosulfates available as an energy source for bacteria to grow; therefore, only species, that can utilize such nutrients, survive, and reproduce [12–15].

In 2009, the Energy Research and Development Institute-Nakornping (ERDI) conducted a study to separate sulfur-oxidizing bacteria from many biofilters operating in livestock farms. Two species that grew well in selective media were identified with a Polymerase Chain Reaction (PCR) technique. The first one, *Bacillus Megaterium*, is heterotrophic bacteria which use organic matter as their carbon source and utilize energy from either chemical reactions or light. The second, *Paracoccus sp.* is a group of autotrophic bacteria which use carbon sources from carbon dioxide and obtain

Table 2.13 H_2S removing performance by different bacteria

Studies	Bacteria	H2S removal efficiency (%)	SV (/h)	Inlet H2S conc. (g/Nm^3)	Biogas flowrate (Nm^3/h)	H2S removal rate ($g/m^3.h$)
[23]	*A. thiooxidans*	91	90	3.04	0.090	370
[33]	*Pseudomonas putida CH11*	92	52	0.18	0.036	9
[34]	*Thiobacillus sp. IW*	80	18	1.52	0.018	22
[35]	*Mixed culture*	83	150	0.32	0.084	46
Barona et al. [32]	*T.thioparus*	95	50	2.34	0.197	111
		100	10	5.86	0.0393	54
	Paracoccus sp.	100	50	2.43	0.197	121.3
		100	10	5.63	0.0393	56.3
	Mixed culture from tofu wastewater	92	50	2.28	0.197	105.3
		99	10	6.41	0.0393	132

energy from sunlight or oxidation of inorganic substances. Since the autotrophic bacteria group can directly use inorganic substances as well as being able to grow under various acidic conditions, this group of bacteria is popular. It was reported that many species in *Paracoccus sp.* such as *Paracoccus Bengalensis sp. nov.* [16, 31] and *Paracoccus Koreensis sp. nov.* are also capable of oxidizing sulfur compounds. ERDI [32] also cultivated mixed cultures from municipal wastewater and tofu-production plant wastewater. The latter performed better than the first in removing H_2S from a biogas stream. Table 2.13 shows the H_2S removing performance by different bacteria.

It was found that *Acidithiobacillus thiooxidans* had the highest removal efficiency. It may be because *A. Thiooxidans* functions in highly acidic conditions (pH 1.8–2.5) which occur when H_2S dissolves in water. However, these acidic conditions may cause corrosion problems to reaction tanks and piping systems. *Paracoccus sp.* had the second best removal efficiency; however, it functions at relatively neutral pH which is less harmful to equipment. Furthermore, it was found that *Paracoccus sp.* grew rapidly to reach the desired working concentration within 3 days compared to 5 or more days required by other species.

2.5 Biofilter Components

A complete biofilter system consists of 9 basic components, a rough schematic is shown in Fig. 2.3. Biogas is driven into the biofilter from the bottom and flows upward through a media bed. If there is more than one biofilter, the columns can be arranged in series. An actual system, located on a livestock farm in Thailand, is shown in Fig. 2.4. These components are described below.

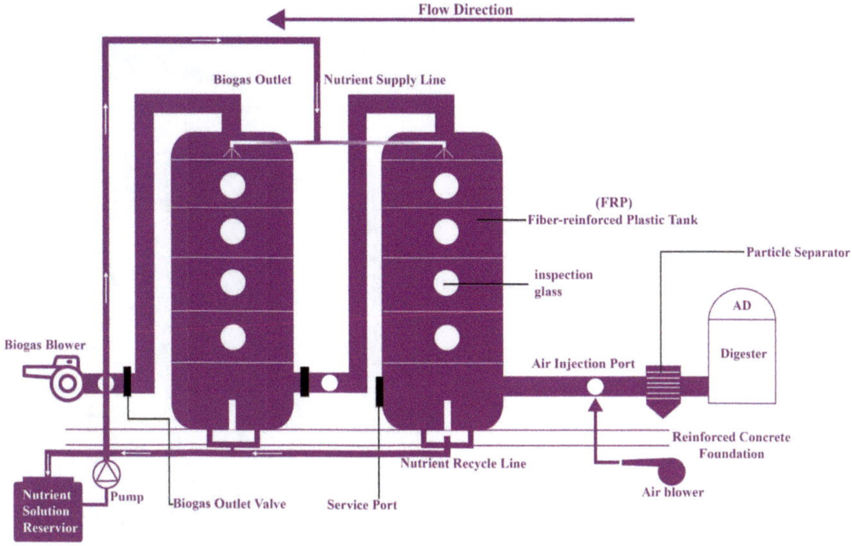

Fig. 2.3 Biofilter schematic diagram

Fig. 2.4 Biofilter installation at a livestock farm in Thailand

1. Particle Separator

A particle separating tank or cyclone is usually made from stainless steel. Its purpose is to trap water and various particles, immediately after the digester and before the biogas enters the pipeline. The principle is to create a centrifugal force or allow gravity to separate heavier particles from a biogas stream (Fig. 2.5).

Fig. 2.5 Particle separator

2. Air Blower

An air blower is used to supply sufficient oxygen to the biofilter to allow the sulfide-oxidizing reaction to occur. As a rule of thumb, the air supply is 2–5% of the biogas flow rate. Air is directly injected into a port at the biogas inlet to induce complete mixing before entering a biofilter column. Too high an airflow rate may lead to a higher amount of nitrogen in the mixed biogas. This is undesirable for some upgrading techniques including membrane separation and water scrubbing which cannot remove nitrogen. Pure oxygen may be used instead of air to avoid such problems (Fig. 2.6).

3. Biofilter Column

A biological filter column is commonly made of Fiber Reinforced Plastic (FRP) which is low cost and able to withstand acidic conditions inside. The diameter can be up to 4 m with the height up to 10 m. The height to diameter ratio is recommended to be 2.5–5 times. At the top of the column, one or more nozzles are equipped to uniformly spray liquid nutrients over the media. The fiberglass tank is often equipped with some transparent side glasses to allow operators to visually observe conditions inside the tank (Fig. 2.7).

Fig. 2.6 Air blower

Fig. 2.7 Biofilter columns

4. Media

Suitable media are packed into a biofilter column to increase the interfacial area between biogas and bacteria. Proper media arrangement allows uniform distribution of sprayed liquid nutrients which promote the growth of bacteria. Plastic media, including Raschig rings and Pall rings, are popular because of their low cost and lightweight. The packing support is located just above the bottom of the column (Fig. 2.8).

5. Strainer

A strainer is made of either a y-shape or u-shape case with a screen inside. It is used to filter any debris from liquid nutrients. The strainer is usually installed at the liquid inlet pipe in order to protect a pump from abrasion and clogging (Fig. 2.9).

Fig. 2.8 Pall rings filled inside the biofilter

Fig. 2.9 Strainer

6. Liquid Nutrient Pump

A centrifugal pump is used to recirculate nutrient liquid which is necessary for the continuous growth of bacteria in a biofilter. The liquid is sprayed over the packed media to keep an adequate moisture condition and provides macro- and micronutrients required by sulfur-oxidizing bacteria. As a rule of thumb, the liquid nutrient flow rate is set at 2% of the total biogas flow rate. After flowing through the media, the remaining liquid is stored at the column bottom and then recirculated back to a liquid nutrient reservoir (Fig. 2.10).

7. Control Panel

Most equipment in the biofilter system are controlled by Programmable Logic Controllers (PLCs) which are housed in a control cabinet (Figs. 2.11 and 2.12).

Fig. 2.10 Centrifugal nutrient pump

Fig. 2.11 Control panel housing

8. Biogas Blower

A centrifugal biogas blower is a vital part of the system. It drives the biogas at a desired flow rate and a pressure which overcomes the total system pressure drop, usually between 5 and 20 kPa. The blower should be located at the very exit of the system, which means the biofilter operates under a slight vacuum (Fig. 2.13).

9. Post Treatment Ponds

Periodically a biofilter requires media washing and cleaning. This operation generates wastewater which is acidic and harmful to the environment. Therefore, a series of post treatment ponds are employed to treat the wastewater in order to meet regulatory standards. These facultative ponds are excavated earth lagoons with 1–2.5 m depth. The hydraulic retention time is between 5 and 30 days depending on the desired effluent quality. In case that land is limited, other high-rate treatment technologies, such as oxidation ponds and activated sludge, might be considered (Fig. 2.14).

Fig. 2.12 Control panels

Fig. 2.13 Biogas blower

2.5.1 *Example of a Biofilter Calculation*

The problem statement is as follows. It is desired to decrease the H_2S level in biogas from 1,600 to 20 ppm. The biogas flow rate is 500 Nm^3/h. A biofilter with *Paracoccus sp.* is chosen. The column size is designed by selecting a Space Velocity (SV) of $50\,h^{-1}$ and a column diameter of 1.5 m. In order to calculate the column size, the following method is used:

The removal efficiency is defined as $\frac{H_2S_{in} - H_2S_{out}}{H_2S_{in}} = \frac{1600-20}{1600} = 98.8\%$

Fig. 2.14 Post treatment pond

1600 ppm of H_2S which means that 1 m^3 of biogas contains 0.0016 m^3 of H_2S. Using the ideal gas law at normal temperature and pressure (0 °C, 101.325 kPa) and a molecular weight of 34 g/mol for H_2S gives a mass of 2.43 g_{H_2S}/Nm^3_{Biogas}

Since 98.8% of this must be removed, every hour the H_2S removal Rate must be $= 2.43\,g/Nm^3 * 500\,Nm^3/h * 0.988/10\,m^3$

$$= 119.91\,g_{H_2S}/m^3/h$$

The media volume is equal to $= \frac{Biogas\ flow\ rate}{Space\ velocity} = \frac{\dot{V}_{BG}}{SV}$

$$= 500/50 = 10\,m^3$$

The retention time equals $= \frac{media\ volume}{Biogas\ flow\ rate} = 10/(500 * 60 * 60) = 72\,s$

The column diameter is set at 1.5 m so the media bed height is:

$$\text{Media Bed Height} = \frac{Media\ Volume}{Cross\text{-}sectional\ Area}$$

$$= \frac{10\,m^3}{0.25\pi\,1.5^2}$$

$$= 5.66\,m$$

The headspace must be designed to allow a solid cone of sprayed liquid nutrients to develop. A spiral nozzle with the spray angle of 120° is used, therefore the nozzle must be located at least a height of $h = \frac{r}{\tan(half\ nozzle\ angle)} = \frac{0.75}{\tan(60°)} = 0.43\,m$ above the media bed. An extra height allowance of 0.30 m for the top and the bottom parts of the column is recommended to accommodate column internals. Therefore,

$$\text{The total height of the column } = 5.66 + 0.43 + 0.30 + 0.30\,\text{m}$$

$$= 6.69 \simeq 7\,\text{m}$$

The diameter to height ratio is then = 1 : 4.67…which is acceptable.

The packing material selected is an off the shelf, commercially available Pall ring made from polypropylene with a nominal diameter of 2 in. which provides a surface area of 102 m^2 per m^3 of media. It has a bulk density of 68 kg/m^3 and the packing factor is 82 m^{-1}.

The biofilter must be supplied with sufficient liquid nutrients to accommodate stable and efficient H2S removal. The liquid flow rate ($\dot{V}_L$ m^3/s) is approximately 2% of the total biogas flow rate or equal to $0.02 \times \frac{500}{60\times60} = 0.0028\,m^3/s$. If the pump power is desired, assuming a pump efficiency of 75% and a desired pressure rise of 145 kPa. The static head and friction loss (H) are assumed to be 15 m.

$$\text{Pump power: } \frac{\dot{V}_L \Delta P}{\eta}$$

$$= \frac{0.0028 \times 145}{0.75}$$

$$= 0.54\,\text{kW}$$

The minimum airflow rate ($\dot{V}_a$ m^3/s) required is approximately 5% of the total biogas flow rate or equal to 0.0069 m^3/s. The air blower must be able to overcome the pressure drop in the column of around 20 kPa and let us assume another efficiency of 75%. The air blower input electrical power can be calculated as

$$\text{Air blower power: } \frac{\dot{V}_a \Delta P}{\eta}$$

$$= \frac{0.0069 \times 20}{0.75}$$

$$= 0.18\,\text{kW}$$

The column design must be ensured that the countercurrent flow rates of biogas and sprayed liquid do not cause any flow irregularities such as flooding or foaming in the column. The flooding point is when all the liquid is carried away by the gas. The flooding velocity is an important design parameter since it establishes the maximum hydrodynamic capacity at which a packed column can operate. It is recommended

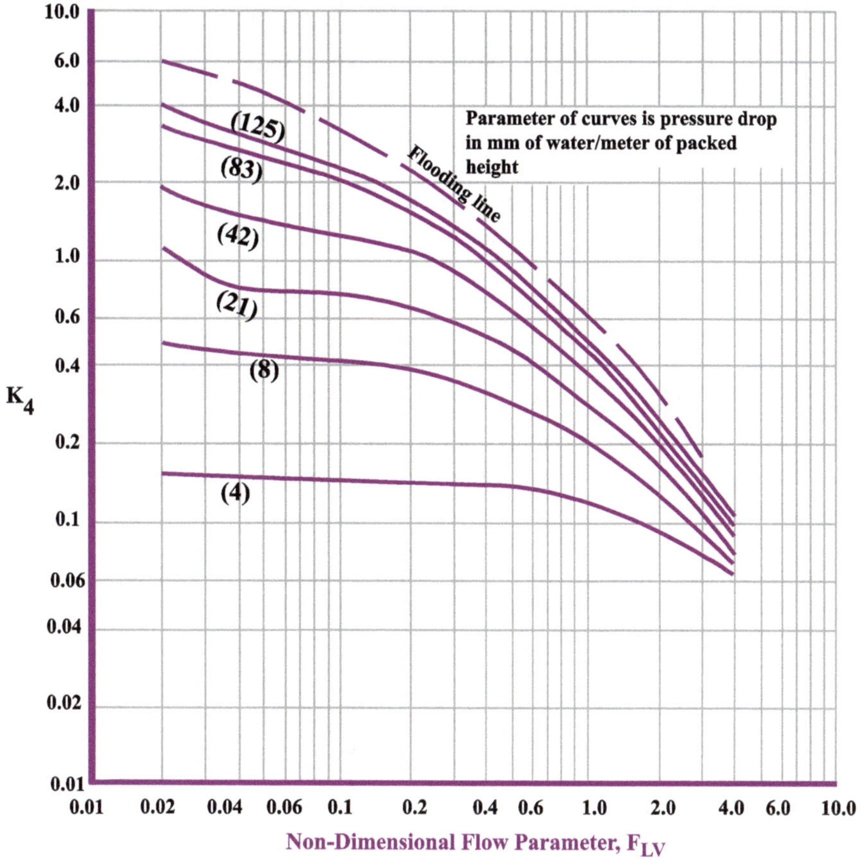

Fig. 2.15 Generalized pressure drop correlation from [36]

that the biogas velocity should not exceed the flooding velocity and the pressure drop should be less than 80 mm of water per meter of media height. Based on the generalized pressure drop correlation chart, Fig. 2.15, the FLV on the x-axis can be calculated as

$$F_{LV} = \frac{\dot{m}_L}{\dot{m}_{BG}} \sqrt{\frac{\rho_{BG}}{\rho_L}} \tag{2.2}$$

where
$\dot{m}_L$= Liquid mass flow rate (kg/s)
$\dot{m}_{BG}$= Biogas mass flow rate (kg/s)
ρ_{BG}= Biogas density (kg/Nm^3)
ρ_L= Liquid density (kg/Nm^3)

$$F_{LV} = \frac{2.8\,\text{kg/s}}{0.159\,\text{kg/s}}\sqrt{\frac{1.15\,\text{kg/Nm}^3}{1000\,\text{kg/Nm}^3}}$$

$$= 0.59$$

At this F_{LV} the K_4 factor, see Fig. 2.15, is read at 1.0.
The term K_4 in Fig. 2.15 is the function

$$K_4 = \frac{13.1(\dot{m}_G^*)^2 F_p \left(\frac{\mu_L}{\rho_L}\right)^{0.1}}{\rho_v(\rho_L - \rho_v)}$$

where
$\dot{m}_G^*$= Gas mass flow rate per column cross-sectional area, (kg/m^{2}s)
F_P= Packing factor, characteristic of the size and type of packing (m^{-1})
μ_L= Liquid viscosity, (Ns/m^2)
$\rho_{L,}\rho_v$= Liquid and vapor densities, (kg/m^3)

Therefore, the flooding velocity, V_W^* which is the value of $\dot{m}_G^*$ where K_4 is the value from the flooding line in Fig. 2.15, can be calculated as

$$V_W^* = \left[\frac{1.0 * 1.15(1000 - 1.15)}{13.1 * 82 * \left(\frac{0.001}{1000}\right)^{0.1}}\right]^{0.5} \tag{2.3}$$

$$= 2.06\,\text{kg/m}^2/\text{s}$$

At the diameter of 1.5 m, the actual gas mass flow rate per column cross-sectional area is: $\dot{m}_G^* = \frac{mass\ flow}{Area} = \frac{0.159}{\frac{\pi}{4}(1.5)^2} = 0.09\,\text{kg/m}^2/\text{s}$ which is less than the flooding velocity, Eq. 2.3. Therefore, the column design is acceptable.

2.6 Biogas Dehumidification

Biogas from the digester is usually saturated with moisture. The quantity of water vapor in the biogas depends on temperature. At higher temperature, the biogas holds a larger amount of moisture. Before sending the biogas through an upgrading system, the moisture must be partially removed in order to prevent condensation. This helps prevent corrosion and equipment malfunction. Moisture also lowers the biomethane heating value, causing problems with certain combustion process. If the gas temperature is lower than the dew point, water vapor condenses and can potentially combine with CO_2, NH_3, and H_2S to form bases or acids, an example of which is shown in Eq. 2.4.

$$H_2S + H_2O \longrightarrow H_2SO_4 \tag{2.4}$$

The maximum allowable moisture varies according to the upgrading techniques. Biogas with saturated moisture is acceptable for water scrubbing but can be detrimental to a membrane module. Hence, a brief discussion on biogas dehumidification as a pretreatment is presented here.

2.6.1 *Dehumidification Method*

Moisture reduction in biogas can be conducted in 3 ways:

(A) Gas heating Heating biogas can reduce relative humidity but it does not remove moisture from the gas. It is used only when there is an abundant and cheap source of heat. This method is mainly used to prevent condensation in pipelines or equipment. However, if the environment temperature decreases unexpectedly, the relative humidity increases and condensation may occur.

(B) Condensation Cooling biogas to the dew point increases the relative humidity to 100%. Decreasing the temperature further induces condensation which can be separated from the gas. It is normally followed by slight heating to keep the gas temperature above the dew point. The cooling is conducted by passing biogas through a refrigerator/chiller. However, the temperature cannot be lowered beyond the freezing point as ice forms and blocks the evaporator gas passage. Thus this method cannot reduce moisture to a very low level. This method is discussed in more detail in Sect. 2.6.2.

(C) Adsorption Moisture adsorption is a dehumidification method by using adsorbents such as silica gel, paper pulp or calcium oxide, etc. A proper design of an adsorption system can provide a very low level of moisture with a water dew point lower than—50 °C. Dehumidification by adsorption is usually employed in an upgrading process or as part of a biomethane post treatment process.

2.6.2 *Vapor Compression System*

Figure 2.16 illustrates a single-stage refrigeration system based on vapor compression principles. A vapor compression system consists of six basic components including:

1. Compressor: Works by compressing the refrigerant vapor from the evaporator to the condenser. The temperature of the vapor increases. The hot pressurized vapor is then passed to a condenser.
2. Condenser: Is also known as a hot coil. It is made of lengthy copper coiled pipes interconnected with fins. It is equipped with a motorized fan which blows air over the outside of the coil. Its main function is to release heat into the environment and

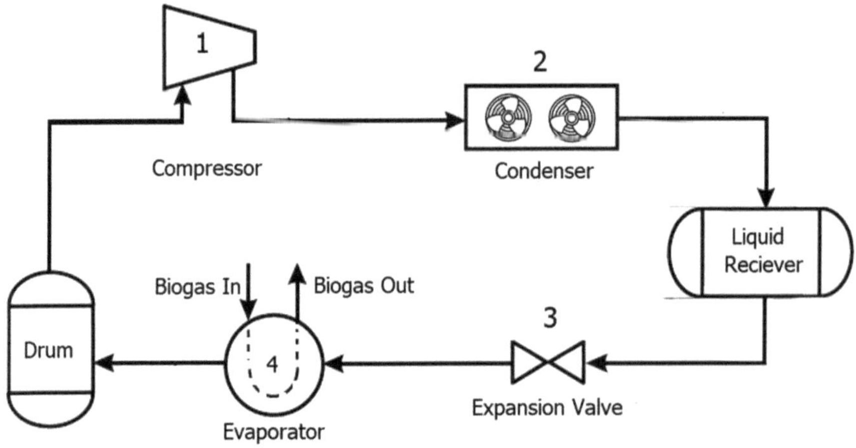

Fig. 2.16 Single-stage biogas dehumidification system

to cool down the high-pressure refrigerant vapor. This results in the condensation of the refrigerant vapor into liquid.

3. Expansion valve: a pressure reducing device. The refrigerant liquid is depressurized to a mixture of liquid and vapor at a low temperature. The low-temperature mixture then flows to an evaporator.
4. Evaporator: also known as a cooling coil. The main function of an evaporator is to receive heat from the substance to be cooled. In this case, the substance is biogas and it is cooled down below its dew point temperature. The moisture is then removed. As the refrigerant is heated, it evaporates before flowing back to the compressor. This completes the working loop of the refrigeration system.
5. Refrigerant: a medium for receiving and transferring heat. This chemical absorbs heat at low temperature and low pressure. It releases heat at high temperature and high pressure. This allows heat exchange between the biogas and refrigerant streams to occur.
6. Another component is a liquid receiver which is a vessel to store the liquid refrigerant after the condenser. It prevents the refrigerant liquid from building up which might lower the surface area of the condenser. The level of liquid in the receiver depends on the load conditions. To guarantee a problem-free operation, the liquid receiver should have at least the same volume as the total refrigerant volume. It is recommended to add an extra 20% volume as a safety factor.

The total system pressure drop in an average chiller is in a range of 30–80 kPa, with the largest pressure drop happening in the condenser. The refrigeration system can also be designed to be a two-stage or three-stage configuration to increase its cooling capacity. However, the capital cost increases as the number of stages increases. The balance between investment and operational performance must be considered.

Table 2.14 Physical properties of selected refrigerants

Physical properties	R-22	R-134a	R-407a
Formula	CHClF2	CH2FCF3	HFC mixture
Molecular weight	86.47	102.03	90.11
Boiling point (1 atm, °C)	−40.70	−26.30	−45.30
Critical pressure (kPa)	4,990	4,059	4,515
Critical temperature (°C)	96.14	101.06	82.26
Critical density (kg/m^3)	523.84	511.90	498.86
Latent heat of vaporization at atmospheric pressure (kJ/kg)	233.75	216.97	238.36
Saturated vapor density at atmospheric pressure (kg/m^3)	4.70	5.25	4.88
Liquid vapor pressure @25 °C (kPa)	1,043.9	665.4	1253.1

Refrigerant Selection Biogas dehumidification relies on commercially available refrigerants such as R-22, R134a, R407a. The required cooling temperature should not be below 0 °C to avoid water freezing. Therefore, commonly used refrigerants are sufficient. The physical properties of selected refrigerants are given below in Table 2.14. Other thermodynamic properties can be obtained from various sources, a good one is the National Refrigerant Reference Guide by National Refrigerant Inc. [37].

2.6.2.1 Design Procedure

Chillers used in biogas upgrading are often designed to be single stage. To design a system, the desired cooling temperature of the biogas must be known. The dew point temperature of biogas leaving the evaporator must be defined. The pressure in the condenser and the evaporator can be determined from a vapor pressure curve of the selected refrigerant.

Design Example Biogas Flow = 500 Nm^3/h or 0.159 kg/s
Inlet Temperature = 30 °C
Outlet Temp = 5 °C
Single-stage refrigeration with R-22 and a Coefficient of Performance (COP) of 2.8
Biogas composition: methane 60% and CO_2 40%
The molar flow rates of both gases can be obtained from the following equations:

$$\dot{n}_{CO_2} M_{CO_2} + \dot{n}_{CH_4} M_{CH_4} = 0.159$$

And

$$\frac{\dot{n}_{CH_4}}{\dot{n}_{CO_2}} = \frac{0.6}{0.4} = 1.5$$

Solving for $\dot{n}_{CO_2}$and $\dot{n}_{CH_4}$ and changing the units to $kmol/h$ gives

$$\dot{n}_{CO_2} = 8.4\,\text{kmol/h}$$

$$\dot{n}_{CH_4} = 12.6\,\text{kmol/h}$$

From steam tables,

$$\bar{h}_{CO_2}\,@35\,°\text{C} = 383.72\,\text{kmol/h}$$

$$\bar{h}_{CO_2}\,@5\,°\text{C} = -696.53\,\text{kmol/h}$$

$$\bar{h}_{CH_4}\,@35\,°\text{C} = 343.58\,\text{kmol/h}$$

$$\bar{h}_{CH_4}\,@5\,°\text{C} = -722.67\,\text{kmol/h}$$

Balancing it all in a steady flow energy balance: $\dot{E}_{in} = \dot{E}_{out}$

$$\dot{n}_{CO_2}\bar{h}_{CO_2}(35\,°\text{C}) + \dot{n}_{CH_4}\bar{h}_{CH_4}(35\,°\text{C}) = \dot{n}_{CO_2}\bar{h}_{CO_2}(5\,°\text{C}) + \dot{n}_{CH_4}\bar{h}_{CH_4}(5\,°\text{C}) + \dot{Q}_{out}(kJ/h)$$

$$8.4(383.72) + 12.6(343.58) = 8.4(-696.53) + 12.6(-722.67) + \dot{Q}_{out}(kJ/h)$$

$$\dot{Q}_{out} = 22508.8\,\text{kJ/h} = 6.25\,kW = 21334.3\,\text{BTU/h} = 0.0213\,\text{mmBTU/h}$$

This is the cooling load to cool biogas (60% methane, 40% CO_2) flowing at $500\,m^3/h$ from 35 °C to 5 °C. The electric power input to the compressor can be calculated from the chiller's COP:

$$COP = \frac{\dot{Q}_{out}}{\dot{P}_{compressor}} = 2.8 \tag{2.5}$$

$$\dot{P}_{compressor} = \frac{\dot{Q}_{out}}{2.8} = \frac{6.25}{2.8} = 2.23\text{ kW} \tag{2.6}$$

Most compressors used for biogas dehumidification are reciprocating compressors which can handle hydrocarbon refrigerants. Other types of compressors are seldom used. Centrifugal compressors are used in a large-scale natural gas processing and are not usually economical at a small scale. Screw compressors are employed only when high discharge pressure is required. It is important to have good mainte-

(a) Chiller unit for biogas pretreatment (b) Evaporator cooling coil unit

Fig. 2.17 Chiller unit for biogas dehumidification

nance procedures for refrigeration systems. The compressor oil should be changed at a frequency that depends on the specifications of the refrigerant, lubricant, and compressor. An oil reclaiming system can be used in large compressors. The oil is trapped at the bottom of a reclaiming drum.

In terms of materials, it is recommended that most piping and components are made of steel. Since the biogas usually contains a certain level of hydrogen sulfide and moisture which are both corrosive, copper or copper-based alloys are not considered suitable materials. The material chosen must be able to withstand a minimum pressure of 1.6 MPa_{gauge} at ambient temperature. For other temperature ranges, the material selection and design should comply with the [38] refrigerant piping code.

Example of biogas dehumidification The dehumidification of biogas by a vapor compression system was demonstrated by [32]. The biogas is pretreated by using a biofilter to remove H_2S before entering the system. A full-scale operation was conducted at a swine farm where a biogas system was located. The size of the system was 4.74 tons of refrigeration and the refrigerant used was R-407C. The temperature of the cooling coil is 10 °C. At the system inlet, the average biogas temperature was 35 °C with an average relative humidity of 90%. After the refrigeration system, the average relative humidity was only 69%. The amount of water condensed from the evaporator was approximately 4.5 L/h while the electrical energy consumption was only 0.02 kWh/m^3 of raw biogas (Fig. 2.17).

References

1. Van Haren M, Fleming R (2005) Electricity and heat production using biogas from the anaerobic digestion of livestock manure - literature review. Technical report, ontario ministry of agriculture, food and rural affairs, Ridgetown College, University of Guelph, Ontario, Canada
2. European Committee for Standardization (2015) Gas infrastructure - Quality of gas - Group H

3. European Committee for Standardization (2017) Natural gas and biomethane for use in transport and biomethane for injection in the natural gas network - part 2: Automotive fuels specification
4. Authur W, Anna L (2006) Biogas upgrading and utilization. Technical report, IEA Bioenergy
5. Hoyer K, Hulteberg C, Svensson M, Jernberg J, Norregard O (2016) Biogas upgrading - technical review. Technical report, ENERGIFORSK
6. European Committee for Standardization (2016) Natural gas and biomethane for use in transport and biomethane for injection in the natural gas network - part 1: Specifications for biomethane for injection in the natural gas network
7. Technical Committee ISO/TC 118 (2010) Compressed air - Part 1: Contaminants and purity classes
8. Electrigaz Technologies (2008) Feasibility study - biogas upgrading and grid injection in the Fraser valley. Technical report, BC Innovation Council, British Columbia
9. Bruser T, Lens P, Truper H (2000) The biological sulfur cycle. In: Environmental technologies to treat sulfur pollution, pp. 47–76. IWA Publishing, London
10. Brock T, Madigan M, Martinko J (2006) Biology of microorganisms. Prentice-Hall, Upper Saddle River
11. Kuenen JG, Robertson LA, Tuovinen OH (1992) The Prokaryotes, chapter The genera Thiobacillus, Thiomicrospira, and Thiosphaera, pp 2636 – 2657. Springer, Berlin
12. Chen G, Geng L, Yan R, Gould D, NG L, Liang T (2004) Isolation and characterization of sulphur-oxidizing thiomonas sp. and its potential application in biological deodorization. Appl Microbiol 39:495–503
13. Kantachote D, Charernjiratrakul W, Noparatnaraporn N, Oda K (2008) Selection of sulfur oxidizing bacterium for sulfide removal in sulfate rich wastewater to enhance biogas production. J Biotechnol 11(2)
14. Wada A, Shoda M, Kubota H, Kobayashi T, Fujimura K, Kuraishi H (1986) Characteristics of H2S oxidizing bacteria inhibiting a peat biofilter. Ferment Technol 64:161–167
15. Sorokin YD, Foti M, Pinkart CH, Muyzer G (2006) Sulfur-oxidizing bacteria in soap lake (Washington State), a meromictic, haloalkaline lake with an unprecedented high sulfide content. Appied Environ Microbiol 73:451–455
16. Ghosh W, Mandal S, Roy P (2005) Paracoccus bengalensis sp. nov., a novel sulfur - oxidizing chemolithoautotroph from the rhizospheric soil of an Indian tropical leguminous plant. Syst Appl Microbiol 29:396–403
17. Hong J, Park K (2005) Compost biofiltration of ammonia gas from bin composting. Bioresour Technol 96:741–745
18. Elias A, Barona A, Rios FJ, Arreguy A, Munguira M, Penas J, Sanz JL (2000) Application of biofiltration to the degradation of hydrogen sulfide in gas effluents. Biodegradation 11(6):423–427
19. Hirai M, Kamamoto M, Yani M, Shoda M (2001) Comparison of the biological H2S removal characteristics among four inorganic packing materials. J Biosci Bioeng 91:396–402
20. Duan H, Koe L, Yan R, Chen X (2006) Biological treatment of H2S using pellet activated carbon as a carrier of microorganisms in a biofilter. Water Res 40(14):2629–2636
21. Lee E, Cho K, Ryu H (2003) Degradation characterization of sulfur containing malodorous gases by acidithiobacillus thiooxidans az11. Korean J. Odor Res. Eng. 2:46–53
22. Ma Y, Yang B, Zhao J (2006) Removal of H2S by thiobacillus denitrificans immobilized on different matrices. Bioresour Technol 97(16):2041–2046
23. Aroca G, Urrutia H, Nunez D, Oyarzun P, Arancibia A, Guerrero K (2007) Comparison on the removal of hydrogen sulfide in biotrickling filters inoculated with thiobacillus thioparus and acidithiobaccillus thiooxidans. Electron J Biotechnol 10(4):514–520
24. Kim JH, Rene ER, Park HS (2008) Biological oxidation of hydrogen sulfide under steady and transient state conditions in an immobilized cell biofilter. Bioresour Technol 99(3):583–588
25. Rattanapan C, Boonsawang P, Kantachote D (2009) Removal of H2S in down-flow GAC biofiltration using sulfide oxidizing bacteria from concentrated latex wastewater. Bioresour Technol 100(1):125–130

26. Chien Chung Y, Huang C, Ping Tseng C (1996) Operation optimization of thiobacillus thioparus ch11 biofilter for hydrogen sulfide removal. J Biotechnol 52:31–38
27. Oyarzun P, Arancibia F, Canales C, Aroca E (2003) Biofiltration of high concentration of hydrogen sulphide using thiobacillus thioparus. Process Biochem 30:165–170
28. Shinabe K, Oketani S, Ochi T, Kanchanatawee S, Matsumura M (2000) Characteristics of hydrogen sulfide removal in a carrier-packed biological deodorization system. Biochem Eng J 5(3):209–217
29. Cho KS, Ryu HW, Lee NY (2000) Biological deodorization of hydrogen sulfide using porous lava as a carrier of thiobacillus thiooxidans. J Biosci Bioeng 90(1):25–31
30. Vlasceanu L, Popa R, Kinkle B (1997) Characterization of thiobacillus thioparus lv43 and its distribution in a chemoautotrophically based groundwater ecosystem. Appied Environ Microbiol 63(8):3123–3127
31. Jun La H, Taek Im W, Ten L, Suk Kang M, Yun Shin D, Taik Lee S (2005) Paracoccus koreensis sp. nov. isolated from anaerobic granules in an upflow anaerobic sludge blanket (UASB) reactor. Int J Syst Evol Microbiol 55(4):1657–1660
32. ERDI (2009) Biogas purification project (in Thai). Technical report, Energy Research and Development Institute, Thailand
33. Chung YC, Huang C, Tseng CP (2001) Biological elimination of H2S and NH3 from waste gases by biofilter packed with immobilized heterotrophic bacteria. Chemosphere 43:1043–1050
34. Park D, Cha J, Ryu H, Lee G, Yu E, Rhee J, Park K (2002) Hydrogen sulfide removal utilizing immobilized thiobacillus sp. iw with ca-alginate bead. Biochem Eng J 11:167–173
35. Barona A, Elias A, Arias R, Cano I, Gonzalez R (2004) Biofilter response to gradual and sudden variations in operating conditions. Biochem Eng J 22(1):25–31
36. Sinnott R, Coulson, Richardson (2005) Chemical engineering design, vol 6, 4th edn. Elsevier Butterworth-Heinemann, Oxford
37. National Refrigerant Reference Guide. National Refrigerant Inc., 6 edn (2016)
38. American Society of Mechanical Engineers (2016) Refrigeration piping and heat transfer components
39. Bahadori A, Vuthaluru HB (2009) Simple methodology for sizing of absorbers for teg (triethylene glycol) gas dehydration systems. Energy 34:1910–1916

Chapter 3
Biogas Upgrading

3.1 Biogas Upgrading to Biomethane

The following processes can be used for carbon dioxide extraction from biogas [1–3]:

- Water absorption or water scrubbing with or without regeneration,
- Pressure swing adsorption (PSA),
- Chemical scrubbing with amine solvent,
- Absorption with chemical solvents,
- Membrane separation,
- Temperature swing absorption, and
- Various other experimental techniques.

Each process is described in more detail in the following subsections.

3.1.1 Water Scrubbing

Water scrubbing uses water to separate carbon dioxide from biogas. In addition to filtering carbon dioxide, this technique can also filter hydrogen sulfide gas at the same time. The process works because these gases are more soluble in water than methane, see Figs. 3.1 and 3.2. The methane molecule is nonpolar so it has difficulty dissolving in polar solvents such as water or alcohol. Manufacturers of water scrubbing plants include DMT, Econet, Greenlane Biogas, and Malmberg Water. This is commercially the most popular separation method with a 40% market share.

It is an entirely physical process. The raw biogas, from the cleaning system, is pressurized and fed to the bottom of a packed column. Water is fed in from the top and exits the column with absorbed hydrogen sulfide and carbon dioxide. Unfortunately, some of the methane also gets absorbed. The wastewater is depressurized in a flash tank. This causes the dissolved gases to separate out. Methane, being less dense, rises to the top where it can be recovered. Since it contains some water vapor this

S. Koonaphapdeelert et al., *Biomethane*, Green Energy and Technology,
https://doi.org/10.1007/978-981-13-8307-6_3

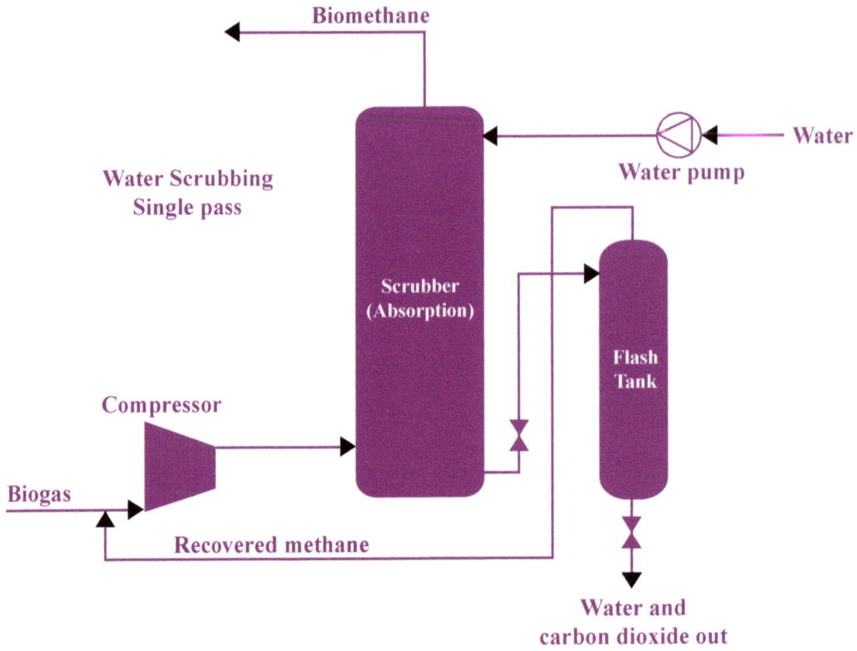

Fig. 3.1 Scrubber for carbon dioxide removal in a non-regenerative wash

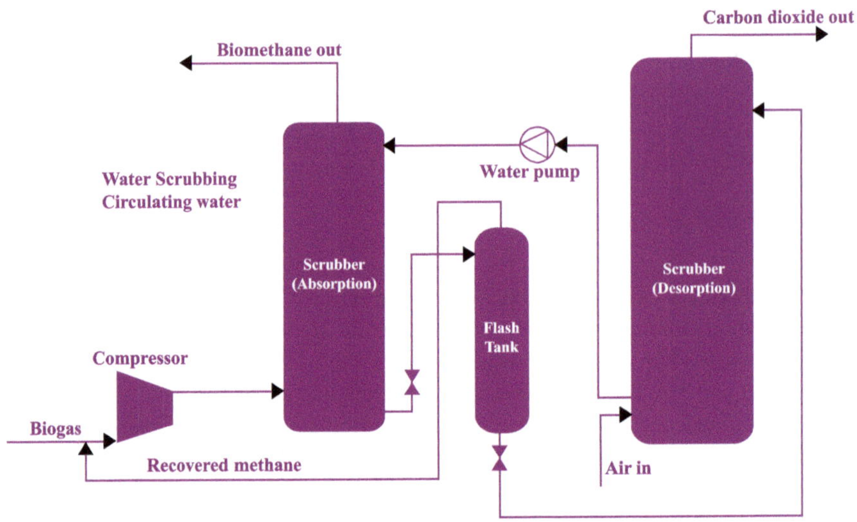

Fig. 3.2 Scrubber for carbon dioxide removal in a regenerative wash

gas is fed back to the start of the cycle and reprocessed. The carbon dioxide that is captured is eventually released back into the environment. The end product has a methane concentration between 88 and 97% by volume. This is sufficient for automotive use [1, 4]. The water flow rate ($\dot{V}_{water}(l/hr)$) can be estimated from the maximum concentration of carbon dioxide possible to dissolve in water, ($c_{CO_2}(M)$). The relationship between these parameters is given by

$$\dot{V}_{water}(l/hr) = \frac{\dot{V}_{CO_2}(gas)(mol/hr)}{c_{CO_2}(M)} \tag{3.1}$$

Modeling this absorption process is challenging. Equation 3.2 can predict the mass transfer flux (N) between the gas and liquid phases but it needs an overall mass transfer coefficient ($K_a(s^{-1})$):

$$N = K_a\,(c - c_{eq}) \tag{3.2}$$

In Eq. 3.2, the concentration driving force is the difference between the concentration of the gas in the bulk phase of the liquid (c (mol m^{-3})) and the dissolved concentration that would be in equilibrium with the bulk phase of the gas (c_{eq} (mol m^{-3})). The overall mass transfer coefficient can be calculated if the mass transfer resistance through both the gas ($1/k_G H_G$) and liquid films ($1/k_L$) are known.

$$\frac{1}{K_a} = \frac{1}{k_G H_G} + \frac{1}{k_L} \tag{3.3}$$

where H_G (kPa) is the Henry's law constant. Henry's law is a gas law that states that the amount of dissolved gas in liquid is proportional to its partial pressure above the liquid. The absorption of both carbon dioxide and methane into water is described by this law.

$$c_A(M) = K_H(M/atm) * P_A(atm) \tag{3.4}$$

In Eq. 3.4, c_A is the concentration of a gas, A, in the liquid phase, K_H is Henry's constant and P_A is the partial pressure of the gas, A. The Henry constant at 25 °C for carbon dioxide is 3.4×10^{-2} M/atm and for methane is 1.3×10^{-3} M/atm, resulting in a solubility for carbon dioxide that is approximately 26 times higher than for methane. For both CO_2 and CH_4, the liquid side mass transfer resistance is dominant over the gas side mass transfer resistance. Many equations have been proposed to calculate the mass transfer coefficients for gas and liquid phases. The correlation from [5] was developed from literature data covering a range of liquids and experimental conditions.

$$k_G = 5.23 \left(\frac{a_p D_G}{RT}\right) \left(\frac{\rho_G u_G}{a_p \mu_G}\right)^{0.7} Sc_G^{1/3} \left(a_p d_p\right)^{-2} \tag{3.5}$$

$$k_L = 0.0051 \left(a_P d_P\right)^{0.4} \left\{\frac{\mu_L g}{\rho_L}\right\}^{(1/3)} \left(Re_L\right)^{2/3} Sc_L^{(-1/2)} \tag{3.6}$$

where

a_p	Interfacial area, m^2/m^3
D_G	Diffusion coefficient, m^2/s
R	Gas constant, J/mol.K
T	Temperature, K
ρ	Density, kg/m^3
u	Superficial velocity, m/s
μ	Dynamic viscosity, kg/m.s
Sc	Schmidt number, [−]
d_p	Particle diameter, m
Re	Reynolds number, [−]
G	Gas phase
L	Liquid phase

Nock et al. [6] used this model coupled with a one-dimensional finite difference approach to calculate the biogas methane concentration at different points along a scrubber. Their results agreed well with experimental data although the experimental data only had two points, the methane concentration at the scrubber inlet and outlet.

The water scrubbing process can be divided into two, based on what happens to the CO_2 saturated water. A non-regenerative process does not recover the water while a regenerative process does. Close to coastal areas, cleaned wastewater from water processing plants provides a near limitless supply of water for use in biomethane scrubbers at virtually no cost. In non-regenerative cycles, a flash tank is used to remove the absorbed methane gas and transport it back to the beginning again as shown in Fig. 3.1. The CO_2 saturated water is discarded. In a regenerative cycle, shown in Fig. 3.2, the absorbed carbon dioxide is separated from the water. This is done in a tank called a desorber. The desorber has a large surface area and operates at a low pressure. Carbon dioxide solubility decreases with pressure. Since a larger quantity of carbon dioxide is dissolved in the water, the composition of the released gas in the flash column will normally be 80–90% carbon dioxide and 10–20% methane. Thereby, the partial pressure of the methane will only be 10–20% of the pressure in the flash column, resulting in a low solubility of methane according to Eq. 3.4. The water that is transported to the desorption column will contain less than 1% of the methane in the raw biogas. An air stream contacts with the water which helps strip the CO_2. Since this process is exothermic, sometimes there is a cooling system at the desorber exit to lower the water temperature. A system producing 150 Nm^3/h of raw biogas needs a water flow rate roughly five times smaller [7]. Care must be taken that the growth of microorganisms is limited. This can be an issue especially in water supplies from sewage processing plants. The microorganisms can grow on pipes and nozzles in the system. It is recommended to change the water periodically to avoid this issue.

Kapdi et al. [4] describe a water scrubbing system that can absorb almost 100% of the CO_2 while the methane lost to absorption is small (<2% by Vol.). The water scrubber was used for both carbon dioxide (CO_2) and hydrogen sulfide (H_2S) removal. The scrubber was 150 mm in diameter and 4500 mm height with a 3500 mm packed bed height. The chamber contains a packed column which increases the absorption area between the carbon dioxide and water molecules. It was designed to enrich the methane content of biogas from 60 to 95%. Pressurized water from the top and pressurized raw biogas from the bottom were fed into the scrubber in a counterflow direction through the packing material, to allow maximum absorption of carbon dioxide in water. The study found that the carbon dioxide was over 10 times more soluble than methane.

The biomethane was stored in a pressure vessel at 20 MPa. The raw biogas production rate was 96 m^3/day from the biogas digester. This daily energy production was equivalent to 1958 MJ/day. Kapdi et al. [8] compared many biogas cleaning technologies in the literature and found that for Indian conditions that water scrubbing was the simplest, continuous, and least expensive method for CO_2 removal. The system's energy requirements were 1000 kJ per 1 m^3 of biogas which is comparable with the energy requirements of other scrubbers [9].

The water scrubbing process can also remove other gases including hydrogen sulfide (H_2S), ammonia (NH_3) as well as suspended solids. The concentration of hydrogen sulfide in biogas can get as high as 3000 ppmv. In general, hydrogen sulfide can be absorbed by water to a concentration of less than 1 ppmv. Since hydrogen sulfide is corrosive and water scrubbing cannot completely eliminate the H_2S, it is common to use a biofilter or an iron sponge filter before the scrubber, see Chap. 2. The accumulation of sulfur along with dissolved hydrogen sulfide can cause an odor especially at the water discharge pipe. Dissolved hydrogen sulfide increases the surface tension of the water. This results in a decrease in the gas-to-liquid contact area. One solution to this problem is to add a chemical for lowering the surface tension at the packing material. In one Swedish biogas plant, a nonhazardous surfactant called kontra spum is used. The biogas capacity is 650 Nm^3/h and only 0.25 L of kontra spum is added to the scrubber per week [10].

The performance of water scrubbers depends on the temperature and pressure inside the scrubber tank as shown in Fig. 3.3. The water solubility of carbon dioxide is linearly proportional to the pressure. However, its solubility decreases with increasing temperature. Effective scrubbers require high pressure and low temperature. The water solubility of CO_2 at 25 °C will have only half the capacity of water at 5 °C. In addition, the properties of the water can affect carbon dioxide absorption. If the water has high alkalinity, this increases the absorption but carbon dioxide absorbed in water causes the mixture to be weakly acidic. Salts or ions can be added to the water, increasing its pH.

Rasi [12] studied biogas upgrading so as to improve fuel quality in transportation applications. An absorption column was used with a height to diameter ratio of 3:1. Water flow rates from 5 to 10 L/min were used with the biogas flowing at rates between

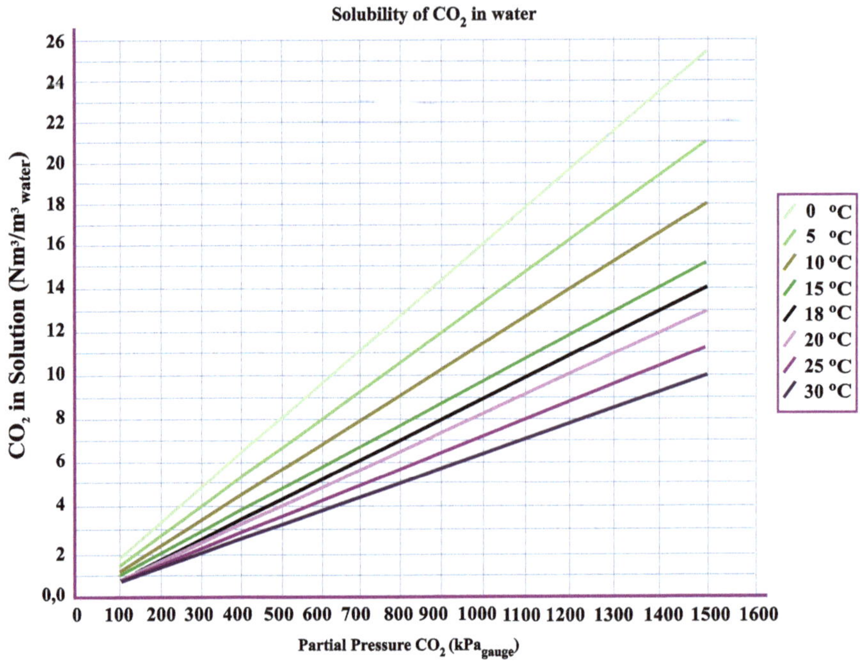

Fig. 3.3 Water solubility of carbon dioxide at various temperatures and pressures from National Institute of Standards and Technology (NIST) [11]

50 and 100 L/min. The pressure was varied between 1000 and 3000 kPa. Figure 3.4 shows the experimental setup. The raw gas had a CH_4 level of $53.2 \pm 1.4\%$ and CO_2 of $40.8 \pm 10\%$ with H_2S of $4.9 \pm 1.2\%$. Increasing pressure resulted in purer methane output. Up to 90% of the CO_2 was removed with H_2S being removed to undetectable levels.

Bansal and Marshall [13] studied the feasibility of recovering waste energy from typical biogas upgrading facilities by means of a hydraulic turbine. They analyzed different types of hydraulic power recovery turbines. The setup is shown in Fig. 3.5. The biogas at high pressure enters the scrubber at the bottom and mixes with a stream of equally high-pressure process water in counterflow in a high-pressure scrubber vessel. The scrubbed wastewater is throttled to a lower pressure, while the purified gas is dried. The authors argue that there seems to be a good potential of recovering this waste energy by expanding this high-pressure fluid in a hydraulic power turbine instead of using a throttling device.

3.1.1.1 Water Quality in Water Scrubbers

Microorganism Growth: It is good practice to treat the water in a water scrubbing system. Especially in the case where the CO_2 saturated water is not recycled (Non-

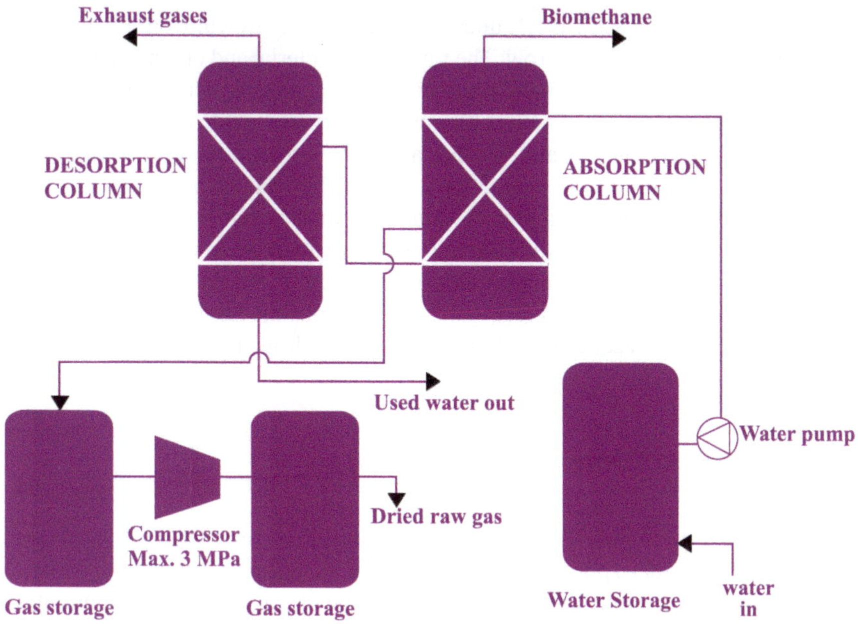

Fig. 3.4 Biogas upgrading using water scrubbing (Adapted with permission from [12])

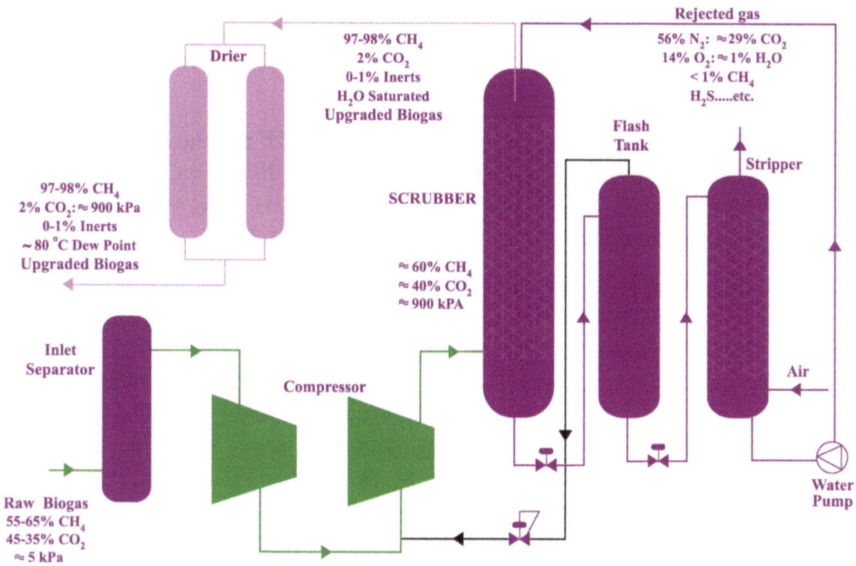

Fig. 3.5 Recovering hydraulic energy in biogas upgrading, adapted with permission from [13]

Regenerative Water Wash). If left untreated the bacteria growth from the nutrition-rich water will exponentially grow. The microbes, bacteria, and organic matter come from the air that passes through the water in the discharge tank. The main cause of the microorganism growth in the tank is because the carbon dioxide is available in large quantities. The environment and weather also play a role in the growth rate. If the water is going to be used for consumption it needs to be cleaned to metropolitan water standards. All fungi, bacteria, and microorganisms must be removed. Facilities where the water is reused, (Regenerative Water Wash) have fewer microorganism problems.

Chemicals can be used to inhibit microbial infection and growth. The operation of the plant must be stopped before using these chemicals. Some common cleaning chemicals include peracetic acid, chlorine dioxide, hydrogen peroxide, and chlorine.

3.1.2 Pressure Swing Adsorption (PSA)

Pressure Swing Adsorption (PSA) is another process to separate carbon dioxide from raw biogas. Biomethane purity levels of above 95% are achievable with this process. It uses adsorbents, such as Zeolites or activated carbon to selectively capture the carbon dioxide. Adsorption is the adhesion of atoms or molecules onto a solid surface. The higher the pressure the more gas is adsorbed. When the pressure is reduced the gas is released or desorbed. Carbon dioxide is attracted to certain adsorbents more strongly than methane. The exact quantity depends specifically on the properties of the adsorbent such as the surface area, composition, and pore size. There are two mechanisms that result in a higher selectivity to CO_2:

1. The first is a stronger surface attachment between CO_2 and the adsorbent.
2. The second is a porous structure that allows CO_2 to diffuse more easily into the adsorbent than CH_4. It helps that the diameter of a CH_4 molecule is 0.4 Å larger. The CO_2 attaches to the surface much faster than the CH_4.

The difference between both types of attachments can be seen in Fig. 3.6 from [14]. The adsorbent is a carbon molecular sieve 3K commercially available form Takeda Corp in Japan. At equilibrium, there is a preferential affinity for CO_2 but the important separation comes from the rapid diffusion of CO_2 as shown in Fig. 3.6b. Equilibrium with CH_4 is only achieved after two days. Several companies develop and commercialize this technology: Carbotech, Acrona, Cirmac, ETW Energietechnik, NeoZeo, Strabag, Mahler, Gasrec, Xebec Inc, and Guild Associates. Commercially, PSA is commonly used due to low energy requirements, safety, flexibility of design, and high efficiency in comparison to other gas separation methods [15]. It has about 20% of the global market share.

It is important to note that no matter what PSA system is selected the hydrogen sulfide should always be removed prior to the columns as it can cause irreversible damage to the adsorbents. The moisture content of the biogas should be sharply reduced for similar reasons.

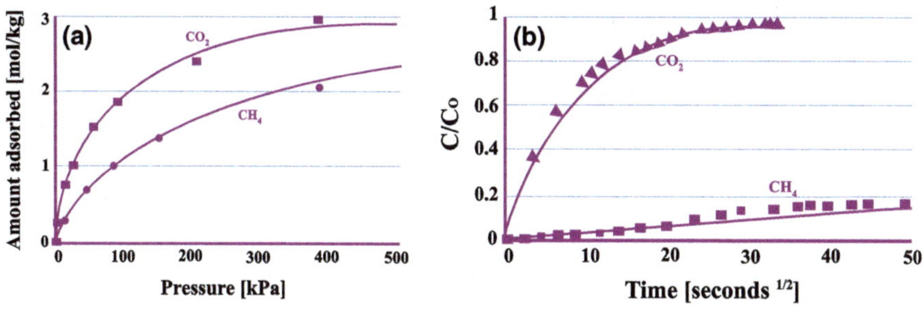

Fig. 3.6 Adsorption of CO2 and CH4 in carbon molecular sieve 3 K at 298 K: **a** adsorption equilibrium; **b** uptake rate curves (Adapted with permission from [14], copyright 2005 American Chemical Society)

After achieving equilibrium, no more adsorption can take place. The adsorbent needs to be regenerated. This is carried out by reducing the pressure and releasing the carbon dioxide. For continuous production, several columns are needed as they will be closed and opened consecutively. One tank is pressurized while the other is depressurized. As the biogas enters the column, CH_4 adsorption is difficult, so it travels through the column quickly. On the other hand, the CO_2 travels very slowly as it is being continuously adsorbed. Once the adsorbent becomes saturated, the CO_2 travels through at the same speed as the CH_4. This is called breakthrough. At this point, the flow should be stopped and the adsorbent regenerated.

Charles Skarstrom in 1960 [16] developed a PSA cycle that uses four cylinders, see Fig. 3.7. This cycle is defined by the following steps:

1. Feed: The raw biogas, with the hydrogen sulfide removed, is fed into a column where the CO_2 is selectively adsorbed.
2. Blowdown: Just before the adsorbent becomes saturated the biogas flow is stopped. The pressure inside the column is reduced causing the CO_2 to be released from the adsorbent.
3. Purge: The CO_2 is purged or removed from the column. This is done using a vacuum and some of the purified methane is fed back to help displace CO_2.
4. Pressurization: A new cycle begins with the raw biogas flowing through the column again at high pressure.

There are four steps in the Skarstrom cycle and therefore if a continuous process is desired, four columns are needed. The process shown in Fig. 3.7 operates at a high pressure of 620 kPa. The blowdown pressure is 310 kPa. The purge happens under vacuum.

The purity of the biomethane at the exit depends on the CO_2 concentration and is given by Eq. 3.7.

$$Purity = \frac{\int_0^{t_{feed}} C_{CH_4} v|_{z=L} dt}{\left(\int_0^{t_{feed}} C_{CH_4} v|_{z=L} dt + \int_0^{t_{feed}} C_{CO_2} v|_{z=L} dt\right)} \tag{3.7}$$

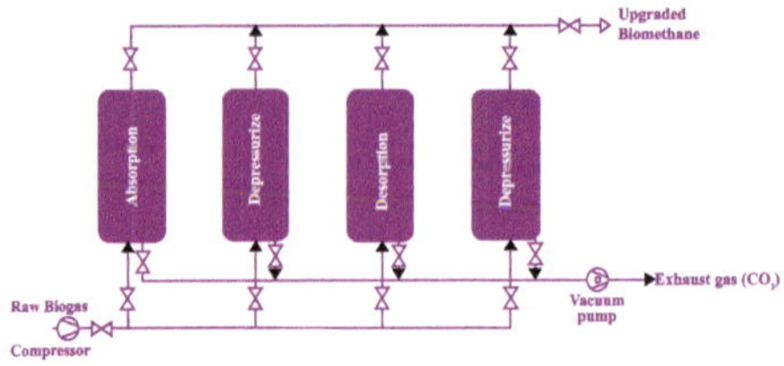

(a) Pressure swing absorption process for carbon dioxide separation from Skarstrom (1960)

(b) PSA system in the central Thailand

Fig. 3.7 Schematic of a pressure swing absorption and example of a PSA plant

where

- C_{CH_4}: The concentration of methane (-)
- C_{CO_2}: The concentration of carbon dioxide (-)
- v: The gas velocity (m/s)
- L: The height of the column (m)
- t_{feed}: The period of time of each cycle where the biogas flows through the column (s)

During the blowdown, some of the adsorbed methane is released with the carbon dioxide. It is important to minimize this loss. The amount of CH_4 lost in the process is termed the CH_4 slip and in the PSA processes is around 3–12% [17]. The methane recovery is the percentage of methane recovered (CH_4 in - CH_4 slip) and is given by Eq. 3.8.

$$Purity = \frac{\int_0^{t_{feed}} C_{CH_4} v|_{z=L} dt - \int_0^{t_{purge}} C_{CH_4} v|_{z=L} dt}{\left(\int_0^{t_{feed}} C_{CH_4} v|_{z=0} dt + \int_0^{t_{press}} C_{CH_4} v|_{z=L} dt\right)} \tag{3.8}$$

A variation on this process is called Temperature Swing Adsorption (TSA) which uses temperature changes instead of pressure to separate the gases. An increase in temperature leads to a decrease in the quantity of CO_2 adsorbed. A modest increase in temperature can cause a relatively large decrease in loading. It is therefore generally possible to desorb any component provided that the temperature is high enough. It is important to ensure that the regeneration temperature does not cause degradation of the adsorbent.

A change in temperature alone is not used in commercial processes because there is no mechanism for removing the CO_2 from the adsorption unit once desorption has occurred. Passage of hot purge gas or steam through the column to purge the CO_2 is almost always used in conjunction with the temperature increase.

TSA is used virtually exclusively for treating feeds with low concentrations of CO_2. The main problem with this technique is the cost. It is a very expensive method and research is needed to bring down the costs.

3.1.3 *Chemical Scrubbing with Amine Solvent*

The capture of carbon dioxide with amine solvents is already widely used in industries such as refineries, petrochemical plants, and natural gas processing plants. This process removes hydrogen sulfide and carbon dioxide. It can be adapted for use with raw biogas. The major obstacles to the large-scale application of this technology are its high energy consumption associated with the CO_2 desorption [18] and amine degradation (up to 30% every year [19]) because of the high stripping temperatures (110–140 °C). This process uses amines such as the most common Monoethanolamine (MEA) or Diethanolamine (DMA), and Diglycolamine (DGA). The acidic CO_2 reacts chemically with the alkaline amine solvent. The capture of carbon dioxide with amine solvents is represented by the following chemical equations [20]:

$$CO_2\,\text{sorption}: \quad RNH_2 + H_2O + CO_2 \longrightarrow RNH_3^+HCO_3^- \tag{3.9}$$

$$CO_2\,\text{desorption}: \quad RNH_3^+HCO_3^- \longrightarrow RNH_2 + H_2O + CO_2 \tag{3.10}$$

The CO_2 reacts reversibly with the amine to form a carbamate. These amines are recycled by increasing the temperature of the solution to 110–140 °C which extracts the carbon dioxide. This is shown in Fig. 3.8. Over time the amine solution becomes rich with CO_2 and is sent to a stripper column. In the stripper, heat is added and used to reverse the chemical equilibrium between the amine and the carbamate, liberating the CO_2. This regeneration is a process that involves high energy consumption. An advantage of chemical scrubbing is that the amine only captures carbon dioxide and does not affect the methane. To reduce the overall

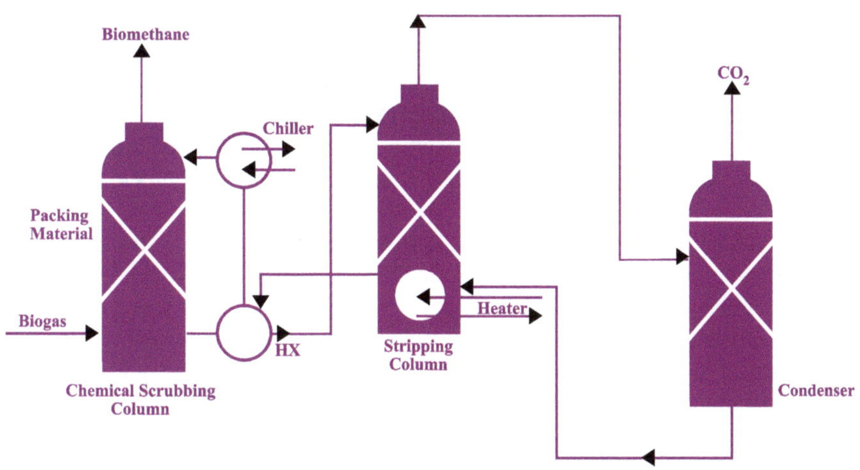

Fig. 3.8 Chemical absorption process for biomethane production

energy requirements the heat released in the sorption reaction can be recycled to heat the desorption reaction. This process can be applied to large-scale gas separation systems. Through a different blend of amines [21] obtained absorption efficiencies up to 96% at lower desorption temperatures, of 90–95 °C. These lower temperatures result in lower ammine degradation, evaporation, and lower equipment corrosion.

Manufacturers of chemical scrubbing plants include Arol Energy, Bilfinger EMS, Cirmac, Energy & waste technologies, Hera, Hitachi Zosen INOVA, MT-Biomethan, and Purac Puregas. Chemical scrubbing has around 20% of the market for biogas upgrading.

3.1.3.1 Other Chemical Methods

There are other chemicals that can react with CO_2 although they are not used very often. One of the oldest methods involves sodium hydroxide (NaOH), Eq. 3.11. It is not used very much anymore because of the high technical requirements to deal with the caustic solution. It can be found in rare applications where very large gas volumes are treated or high concentrations of H_2S are present [22].

$$NaOH + CO_2 \longleftrightarrow NaHCO_3 \tag{3.11}$$

Calcium hydroxide (CaOH) is a cheap and environmentally friend solvent but suffers from operational problems such as forming a hard coating on the column surface. It also causes chemical skin burns. Limewater is a medium-strength base made from aqueous solutions of calcium hydroxide. Limewater turns milky in the presence of carbon dioxide due to formation of calcium carbonate [3]:

$$Ca(OH)_2 + CO_2 \longleftrightarrow CaCO_3 + H_2O \tag{3.12}$$

However, this reaction is not used in industry because of the low rates of absorption of CO_2. Although researchers such as [23] are trying to increase the solubility of CO_2 with the use of Microbubble Generators (MBG). These produce large quantities of bubbles in micrometer sizes. The resulting high contact surface area increases the absorption of CO_2 but more progress is needed for this to compete with other processes.

Tippayawong and Thanompongchart [24] studied the chemical absorption of CO_2 and H_2S by solutions of sodium hydroxide (NaOH), calcium hydroxide ($Ca(OH)_2$), and Monoethanolamine (MEA), in a packed column. The biogas was circulated in a countercurrent flow. Over 90% CO_2 removal efficiency was achieved and the H_2S was removed to below the detection limit. The absorption capability was however transient. Saturation was reached in about 50 min for $Ca(OH)_2$, and 100 min for both NaOH and MEA.

Diao et al. [25] studied the removal of carbon dioxide by ammonia absorption. The carbon dioxide reacts with ammonia at different temperatures. Under dry conditions, room temperature, and a pressure of 1 atm, the NH_3 reacts with CO_2 to produce ammonium carbonate (NH_2COONH_4). Initial carbon dioxide concentrations of 10% (v/v) and an ammonia concentration of 0.14 mol/L effectively filtered 95–99% of the carbon dioxide at a temperature of 35 °C. The process is however highly dependent on the temperature.

3.1.4 Absorption with Physical Solvents

The method is similar to water scrubbing, except that organic solvents are used to absorb CO_2 from biogas. This process is most commonly used in the United States. It is a physical process used in industrial gas separation. These are physical solvents, unlike the amine solvents which are chemical solvents that rely on a chemical reaction. It is a robust technology that can handle various impurities in the biogas. It captures any acidic gas and therefore can capture other impurities such as H_2S, NH_3, and VOCs as well as CO_2. The process is not harmed or compromised by high concentrations of H_2S or NH_3 in the raw gas. The solubility of CO_2 in the physical solvent can be five times higher compared to water. This results in a smaller column than a water scrubber since less solvent is required. Additional drying of the upgraded biogas is not required because moisture is easily absorbed. However, the regeneration of these organic solvents is difficult due to the high solubility of CO_2. Oxygen and nitrogen present in the raw gas will, however, pass through the physical scrubber and will be present in the biomethane product.

The Selexol process (which is licensed by Honeywell Universal Oil Products), utilizes a proprietary solvent (a mixture of the dimethyl ethers of polyethylene glycol) which absorbs acidic gases from the biogas at relatively high pressures, between 2 and 14 MPa. The rich solvent is then reduced in pressure to recover the gases. It can operate selectively to recover hydrogen sulfide and carbon dioxide as separate streams. Genosorb is another solvent used in this process. It is also a mixture of dimethyl ethers and polyethylene glycol. Manufacturers of physical solvent plants include BMF HAASE Energietechnik and Schwelm Anlagentechnik. In spite of the benefits of this technology it only has 6% of the biogas upgrading market [26] (Fig. 3.9).

Gomez-Diaz et al. [27] studied the relationship between the gas/liquid mass transfer process of carbon dioxide in an aqueous solution of glucose and glucosamine. The studies were carried out in a cylindrical bubble column, see Fig. 3.10. The glucose and glucosamine were combined in solution at a concentration of 0–0.4 M and the carbon dioxide flowed at rates of 18, 30 and 40 L/h. The study found the capture of carbon dioxide is dependent on the flow rate and solution concentration.

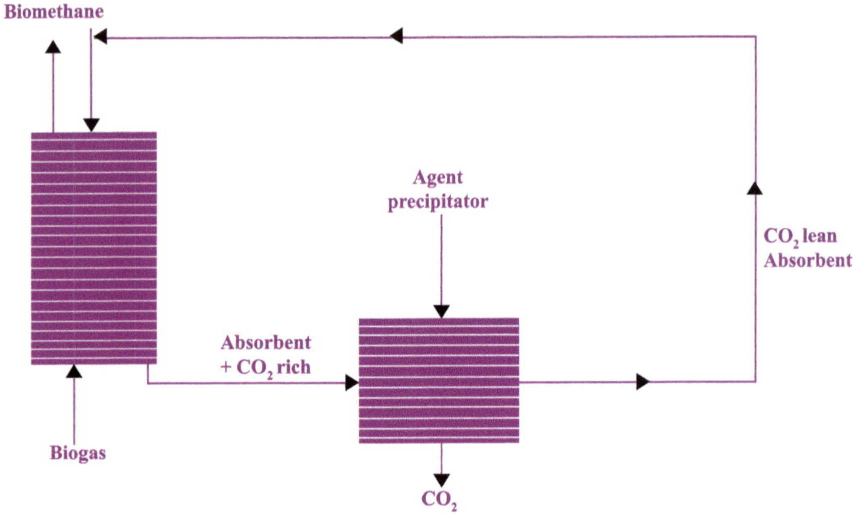

Fig. 3.9 Physical solvent absorption process

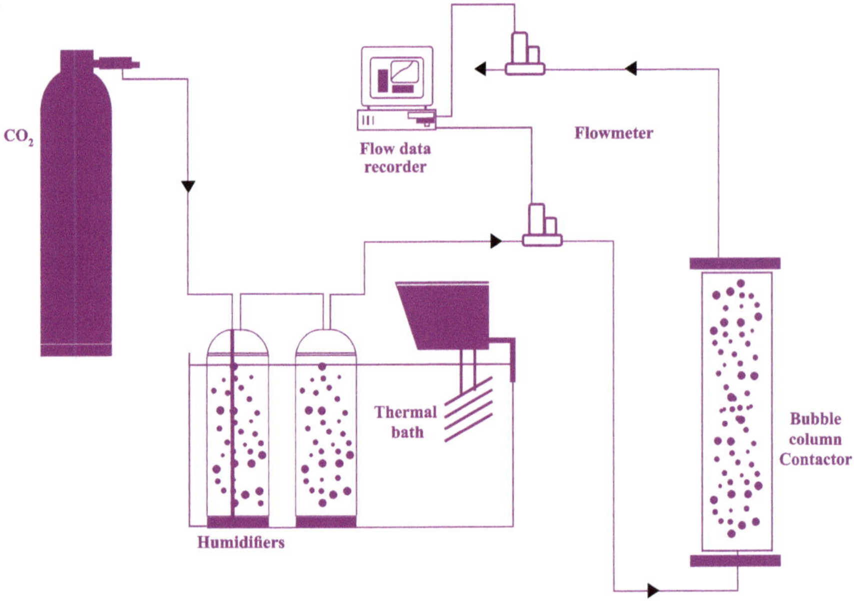

Fig. 3.10 Carbon dioxide capture with a bubble column (Adapted with permission from [27])

3.1.5 Membrane Separation

Membrane separation is another process used to upgrade biogas. It uses the fact that gases have different permeabilities through a membrane fiber. Generally, nonporous membranes made from polyaramide, polyamide, and polyimide are used for gas separation. The membrane structure is asymmetric with a selective layer located on top of a porous supporting layer. Sometimes this selective layer is coated with silicon materials to seal off defects that may have occurred in the manufacturing process. The membrane separation commonly occurs at pressures in the range of 1–4 MPa. This results in higher pressures in the produced biomethane than for other upgrading techniques. Sometimes this is not an issue but in certain end applications, the biomethane pressure will need to be reduced.

Membranes have a pore size of less than 1 nm to separate the gases. Carbon dioxide and hydrogen sulfide diffuse at higher rates than methane. Over time methane becomes concentrated on one side of the membrane. Typically there would be several membranes in series. Most membranes are sensitive to liquid water, oil, and particles and these need to be removed prior to membrane contact. Condensation on the surface of the membrane should be avoided especially if contaminants such as H_2S or NH_3 are present in the gas. If they combine with the moisture, acid is formed on the membrane surface. Moderate concentrations of H_2S do not damage the membrane surface as long as there is no water condensation. VOCs are commonly removed prior to the membrane as certain VOCs can permanently damage the membrane.

Biomethane with a methane concentration of up to 96% is achievable with this method. Few consumables are used in a membrane upgrading plant which distinguishes it from other upgrading plants. There are membrane upgrading plants on the market which have been running successfully with their initial membranes for more than 10 years [28].

A high-pressure operation results in gases being present on each side of the membrane as shown in Fig. 3.11. If one side of the membrane has an absorbing liquid, to flush the CO_2 then operation at a pressure close to atmosphere is possible, see Fig. 3.12. The trade-off for membrane separation is biomethane purity v methane

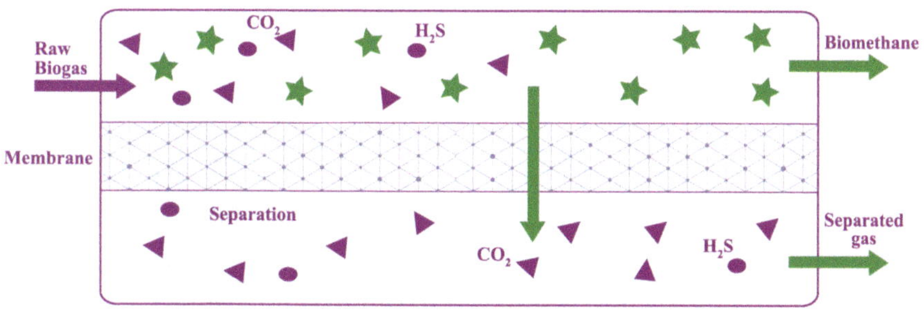

Fig. 3.11 Gas distribution through a membrane

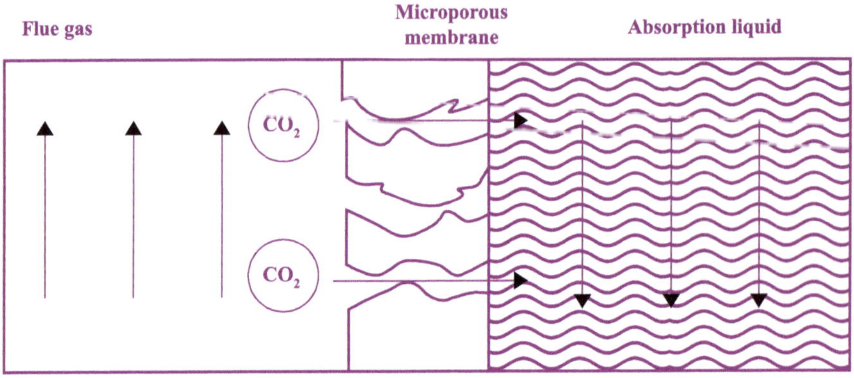

Fig. 3.12 Membrane separation of carbon dioxide with an aqueous solution

slip (losses). The higher the biomethane purity, the higher the methane losses in the output stream. Sometimes the permeated gas is recirculated to recover some of this methane.

For dense membranes, the solution-diffusion model is the most widely accepted model to explain the movement of gas molecules through the membrane layer. The solute flux is the product of the diffusion coefficient and concentration gradient across the membrane thickness. The process is governed according to Fick's law. Figure 3.13 shows diffusion of gas as a result of a concentration gradient. The concentration at the surface is equal to the partial pressure of the gas times its sorption coefficient, $c_i = S_i P_i$.

$$J_i = -D_i \frac{dc_i}{dx} = \frac{D_i S_i (P_{ih} - P_{il})}{\delta_m} \tag{3.13}$$

J_i = Flux ($m^3_{STP}/m^2.s$)
c_i = Concentration (m^3_{STP}/m^3)
D_i = Diffusion coefficient (m^2/s)

Fig. 3.13 Typical gas separation membrane structure

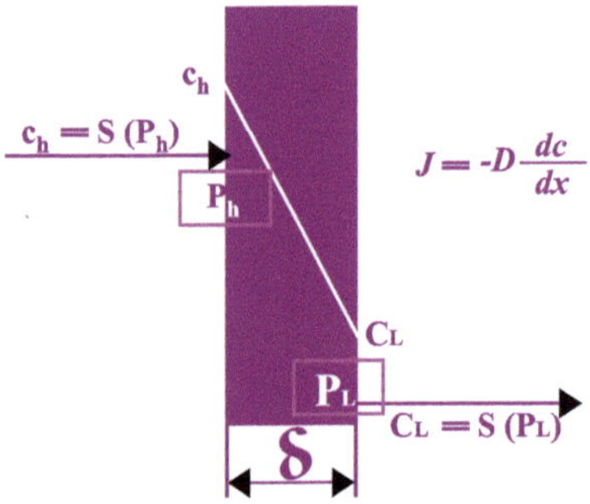

δ_m = Membrane thickness (m)
S_i = Sorption coefficient ($m^3_{STP}/m^3.Pa$)
P_i = Partial pressure (Pa)

The product of D_i and S_i can be defined as membrane permeability ($\bar{P}_i$) in the unit of Barrer, one unit equals $10^{-10}cm^3_{STP}/cm.s.(cmHg)$. The permeability of CO_2 is generally higher than that of CH_4 in all types of materials. The permeability is not as important as the preference of the membrane has to CO_2 over CH_4. So, although the gas permeability of rubbery membranes is significantly higher than those of glassy membranes, their CO_2/CH_4 selectivity is lower. Therefore, membranes made from glassy polymers are usually selected for gas separation to minimize the number of membrane stages. The CO_2/CH_4 selectivity of a typical polyimide is $\frac{\bar{P}_{CO_2}}{\bar{P}_{CH_4}} = \frac{13}{0.4} = 32.5$ times, indicating the relatively fast permeation of CO_2 compared to CH_4. For this reason, it is often used as material of choice for biogas upgrading membrane.

New classes of polymers are being developed for biogas separation. They include Thermally Rearranged (TR) polymers, Polymers of Intrinsic Microporosity (PIMs), Room-Temperature Ionic Liquids (RTILs), perfluoropolymers, and high-performance polyimides. TR polymers in particular show selectivity values well above 32, see Fig. 3.14, which favors them for use in future commercial membrane separators.

There are many commercial membrane companies including, MEDAL, DuPont, Evonik, Ube, Air Liquide, Arol Energy, Axiom Angewandte Prozesstechnik, Bebra-Biogas, Cirmac, DMT, Eisenmann, EnviTec Biogas, Gastechnik Himmel Hitachi

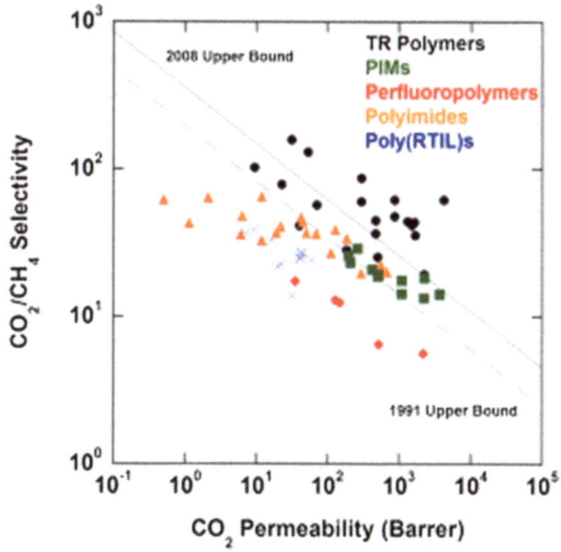

Fig. 3.14 CO_2/CH_4 separation properties of various emerging polymer materials reported in the literature (Reprinted with permission from [29])

Zosen INOVA, and Pentair Haffmans. The materials used are proprietary but often based on polyimide or polyamide. This technology has around 10% market share.

3.2 Economic Analysis

The economic cost of producing biomethane is divided into three sections: the cost of biogas production, cost of biogas upgrading, and additional costs. These are all outlined in the following subsections.

3.2.1 Biogas Cost

ERDI [30] estimated the costs for biogas production from farms that use anaerobic reactors in Southeast Asia and Europe. The analysis was done for lagoon reactors and continuously stirred anaerobic reactor systems. A lagoon is a passive plant which captures the biogas under an impermeable cover. A tank is a digester where the wastewater is more actively controlled by continuously or intermittently heating and mixing. In general, costs can be divided into capital costs which include the cost of designing and building the plant, and operating costs which include mainly the cost of purchasing electricity and raw materials for running the plant. These costs will vary in different parts of the world. The construction and operating costs per unit of biogas produced in Europe and SE Asia are shown in Tables 3.1 and 3.2. The assumptions that are included in this cost analysis are:

- There is no revenue for the treatment of waste from the agricultural sector.
- There is no revenue from the sale of carbon credits.
- There is no revenue generated from selling the nutrients, at the end of the process, as fertilizer.
- There are no expenses relating to the separation of manure and compost but there is also no revenue from the sale of compost.
- A biogas plant has a typical lifetime of 15 years.
- The plant has a capacity factor of 80%.
- The financial interest rate is 7%.
- The specific energy of raw biogas is 21.6 MJ/Nm^3.
- The specific energy of the finished biomethane is 34.8 MJ/Nm^3.
- The specific energy pure methane is 39.6 MJ/Nm^3.

Combining these cost ranges, for two biogas production rates (400 Nm^3/hr and 1000 Nm^3/hr) in Europe and SE Asia yields the graph shown in Fig. 3.15. This cost assumes that the raw material for the plant is free. Energy crops are not used. It shall be shown later that paying for these crops significantly adds to the biogas cost especially in Europe. The first three data points are for production capacities of 400 Nm^3/h with the second three being 1000 Nm^3/h. Interestingly the cost of biogas

Table 3.1 Biogas plant construction costs in SE Asia and Europe, from [30]

Expense	SE Asia		Europe	
	Range 0–500 Nm^3/h	Range 500–1000 Nm^3/h	Range 0–500 Nm^3/h	Range 500–1000 Nm^3/h
Construction ($)	Lagoon: $2–2.5 million	Lagoon: $3.5–4.2 million	Lagoon: N/A	Lagoon: N/A
	Tank: $2.5–3.2 million	Tank: $4.5–5.5 million	Tank: $3.5–5.0 million	Tank: $5.5–8.5 million

Table 3.2 Biogas plant operation costs in SE Asia and Europe, from [30]

Expense	SE Asia ($)	Europe ($)
Electricity ($/$Nm^3$) Thailand electricity cost at 4 THB/kWh or $0.13/kWh	Lagoon: $0.005–0.010/$Nm^3$	Lagoon: $0.01–0.02/$Nm^3$
	Tank: $0.005–0.010/$Nm^3$	Tank: $0.01–0.02/$Nm^3$
Raw material ($/$Nm^3$)	Wastewater/solid waste—no cost	Lagoon: N/A
	Energy crop: $0.10–0.30/$Nm^3$	Energy crop: $0.25–0.45/$Nm^3$
Labor $/$Nm^3$	$0.010–0.015/$Nm^3$	$0.020–0.030/$Nm^3$
Maintenance $/$Nm^3$	Lagoon: $0.010–0.020/$Nm^3$	Lagoon: N/A
	Tank: $0.015–0.025/$Nm^3$	Tank: $0.015–0.025/$Nm^3$

produced in a tank is comparable in both regions. The European biogas from tanks costs about 80% more compared with lagoons in Southeast Asia. The cheapest costs come from a high production rate from lagoons in SE Asia at $0.08/$Nm^3$of which 68% comes from the capital cost of construction. The most expensive being the low production rate from Europe at $0.27/$Nm^3$ of which 72% was the capital cost.

The situation is radically different if energy crops are purchased. The most expensive price in Europe increases to $0.72/$Nm^3$, a huge 62% of which comes from the purchase of the digester feedstock.

Changing the units of cost to $\$/GJ_{biogas}$ gives, a price for a high volume flow rate lagoon in SE Asia, of $\$\,3.7/GJ_{biogas}$. The most expensive production cost of biogas, a low volume flow European tank, works out at $\$\,12.5/GJ_{biogas}$. Compare these to the residential price of natural gas in the US of around $\$\,8.47/GJ_{gas}$ and an average European price (2017) of $\$\,18.17/GJ_{gas}$ with a Japanese residential price of $\$\,29.25/GJ_{gas}$. The price for Europe varies from country to country, this number here is an average. It is worth noting that gas residential prices are typically higher than the prices of natural gas used in industry and commerce. These prices provide an upper bound on the economics of biogas production. These numbers suggest that raw biogas is cost competitive at the production stage. The biogas yield is a critical factor in this process. However if the NG price falls, this would render biogas producers uncompetitive without feed in tariffs or other such government subsidy.

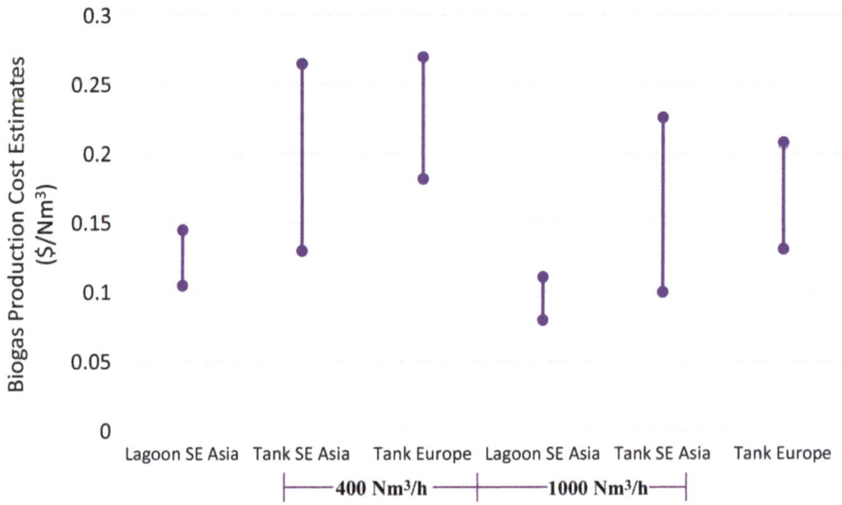

Fig. 3.15 Range of biogas production costs in SE Asia and Europe for different biogas plants

If energy crop prices are included, the European biogas cost jumps to $\$\,33.3/GJ_{biogas}$ which is already uncompetitive. The raw biogas needs to be upgraded to biomethane for grid injection, for use as heating in residences or as a transportation fuel. Upgrading adds costs which are presented in the next section.

3.2.2 *Cost of Biogas Upgrading*

There are many different upgrading processes as described previously throughout this chapter. The data for these upgrading costs are taken from [26]. These are shown in Table 3.3. They are presented here in US dollars but come primarily from European processing plants. A high biomethane production rate lowers the capital cost per unit of biomethane produced. Upgrading costs, similarly to biogas costs, include:

- Land prices.
- Design and construction of plant.
- Control and measurement equipment. Gas quality measurement, pipeline, and storage construction.
- Odorization of the biogas as discussed in Sect. 3.3.2.
- Operations and maintenance including labor, chemical handling, and disposal of chemicals.
- Expenses for hydrogen sulfide gas removal.

There are also external expenses for handling and cleaning of water, air, and residue and returning them to the environment in a clean and safe way.

Table 3.3 Average cost of upgrading to biomethane for different upgrading processes. (Reprinted with permission from [26])

Technology	Biomethane flow rates(Nm^3/h)	Capital costs ($)	Operating costs (electricity cost at $0.13/kWh) ($\$/Nm^3$)
Water scrubbing	100–500	$0.627–1.425 million	$\$0.026-0.039/Nm^3$
	1000–	$2.28–2.05 million	
Physical absorption	250	$1.28 million	$\$0.026-0.065/Nm^3$
	1000	$1.7 million	
Chemical absorption	600	$2.19 million	$\$0.065-0.091/Nm^3$
	1800	$3.06 million	
Pressure swing absorption	600	$1.85 million	$\$0.031-0.078/Nm^3$
	2000	$3.4 million	
Membrane	100–400	$0.684–1.14 million	$\$0.026-0.052/Nm^3$
	1000	$2.28 million	
Cryogenic		n/a	$\$0.052-0.13/Nm^3$

Biomethane operating expenses depend on labor costs and industrial electricity prices. In America, the average price of electricity is around 13 cents per kilowatt hour with European prices ranging from 15 to 35 cents per kilowatt hour. A way to standardize the cost is to use units such as the energy consumed per unit volume of biomethane production. However, the units used here are $\$/Nm^3$ to keep it consistent with the biogas costs. In that way, the costs can be easily compared and added.

The upgrading capital cost data, from two sources [26, 31], are plotted in Fig. 3.16. The costs are only representative as they do not include planning, permission, construction or any costs specific to a location such as permits or licenses. The upgrading technologies broadly follow a similar cost curve. Both the cost and the trend line is similar for both curves. There are substantial savings to be made if production at a rate of 1000 Nm^3/h can be achieved. Cucchiella et al. [32] estimated that a biomethane production rate of at least 350 Nm^3/h was needed to be profitable in Italy. Going beyond this rate the capital costs per unit biomethane produced tend to level out. As technology and management experience increases these costs can be reduced.

The costs presented in Table 3.3 should be taken as an estimate that will vary depending on the plant location. Also included in the final row is the operating cost for a cryogenic separation process. There are only a handful of these plants globally and capital cost data was not available. Cryogenic cooling is not yet a cost-effective upgrading technology with only a small (0.4%) market share. It will be discussed in more detail in Chap. 7. Using similar assumptions to the biogas analysis above, 7% interest rate, 15-year lifespan, 80% capacity factor the total cost of upgrading for the different technologies can be estimated per cubic meter of biomethane produced, Fig. 3.17. The error bars correspond to the range of operating costs.

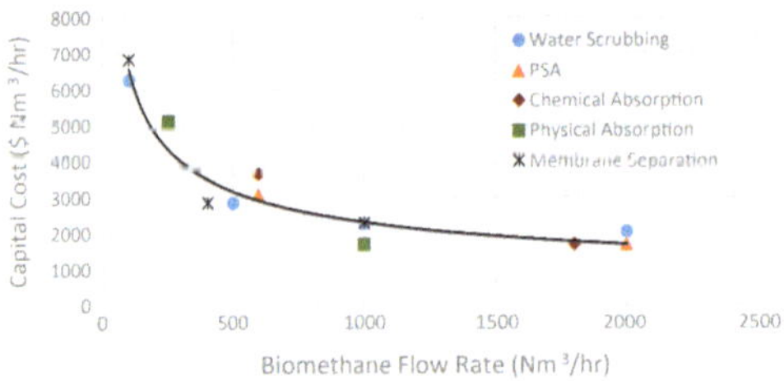

(a) Specific capital costs for biomethane upgrading processes (from Kapoor et al. (2019))

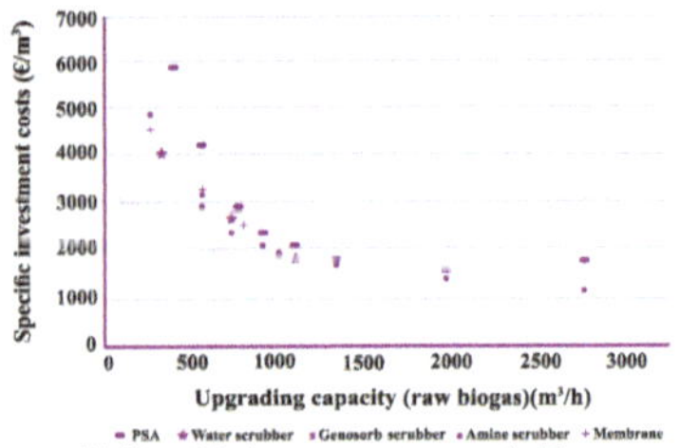

(b) Specific investment costs in Euros per Nm^3 raw gas upgrading capacity of five biogas upgrading methods (Reprinted with permission from Beil and Beyrich [31])

Fig. 3.16 Specific capital costs for biomethane upgrading processes from two separate sources

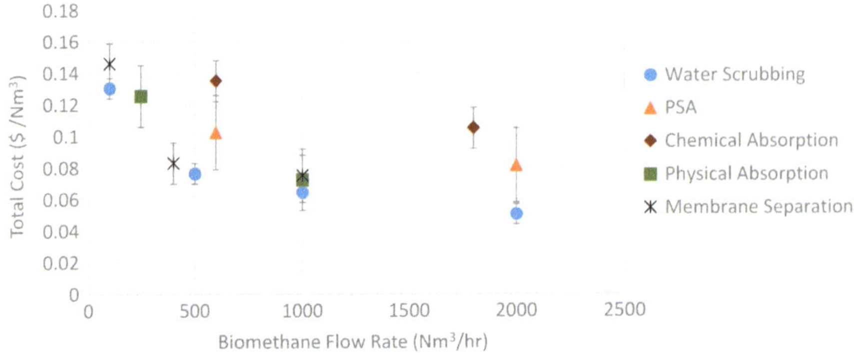

Fig. 3.17 Total cost of upgrading to biomethane for five technologies

At a production rate of 400 Nm^3/h, the average cost of upgrading to biomethane using membrane technologies is $ 0.083/Nm^3 which works out at $ 3.84/GJ divided between a capital cost of $ 2.03/GJ and an operating cost of $ 1.81/GJ. This cost drops to $ 3.47/GJ at a flow rate of 1000 Nm^3/h.

3.2.3 *Total Biomethane Production Cost*

The total biomethane production cost is the total of the biogas cost and the upgrading cost. This is the cost of producing biomethane and can be compared with natural gas prices to see if it is economical or not. There is a wide range of biogas and biomethane costs so for this comparison the average values, across all flow rates, in Europe and SE Asia will be taken. The total production cost can be classified into:

1. Average European biogas production cost (including energy crops) $ 25.4/GJ.
2. Average European biogas production cost (excluding energy crops) $ 9.15/GJ.

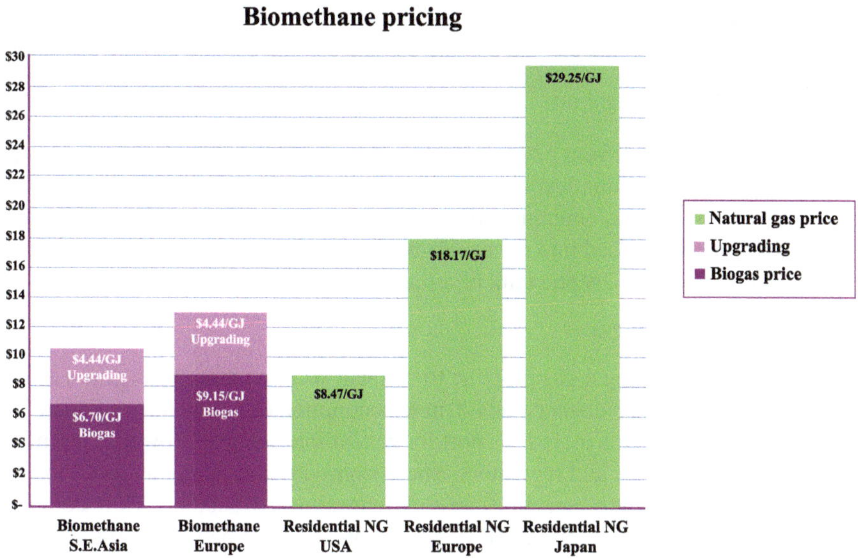

Fig. 3.18 Cost of biomethane production, compared to the price of natural gas

3. Average SE Asia biogas production cost $ 6.7/GJ.
4. Average SE Asia upgrading biogas to biomethane cost $ 4.44/GJ.

This gives a total cost (excluding energy crops) in Europe of approximately $ 13.59/GJ and in SE Asia of $ 11.14/GJ. These are just the costs, they do not include any profits for the plant operators. These costs are compared with the residential price of NG in Fig. 3.18. The reason why biogas is not produced widely in the US is from the cheap and abundant supply of natural gas available. Southeast Asia can produce at a competitive rate so long as the biogas substrate is free. Europe is also cost competitive, assuming free wastewater, but the risks are higher. A fall in gas prices or sudden increase in costs, for example, electricity prices could change the economic case. The issue is in finding markets and uses for biomethane in the absence of a natural gas grid.

The outlook may be more positive than this analysis indicates. This analysis does not include gate fees. These are fees paid to the biomethane plant operator by those that wish to dispose of their organic waste. In many countries the government subsidies and provides incentives to produce biomethane. In addition, biogas plants can have alternative revenue streams from selling compost and carbon credits as well as being more environmentally friendly than traditional waste disposal methods. These extra revenues can raise money to contribute to plant profitability.

This comparison in Fig. 3.18 does not include transportation fees, taxes or fees charged by gas grid operators.

3.2.4 *Other Costs*

(A) Waste Stream Mitigation

The gas waste products from a water scrubbing system need to be disposed of properly. The allowed emission level depends on the local air quality and emission regulations. If the waste gas contains more than 10% methane, it should be combusted instead of directly released into the environment. The wastewater should be cleaned with a treatment system, septic tank or a reactor.

(B) Gas Grid Connection

This is the cost of laying a gas pipeline to connect with a national gas grid. Laying underground pipelines a distance of 400 m to accommodate a flow rate of 240 m^3/h, costs about $90,000. This includes a port for gas sample analysis, flow measurement instrumentation, valves, and regulators. Specific gravity measurements, an odorizer unit, and a short pipeline connection can cost approximately $60,000. More complex systems, including gas injection and compression, can cost between $100,000 and $400,000 which would not be suitable for the scale of a small farm. More details on pipeline design are given in Chap. 6.

(C) Pressurizing Cost

Biomethane distribution inside a national gas pipeline needs to be pressurized to the pipeline pressure. Table 3.4 shows the cost of pressurizing the biomethane, from an initial pressure from the upgrading plant to a final pipeline pressure of 3300 kPa. These costs do not include labor costs. In some cases, higher pressures are needed. In a biomethane fueled engine operating at 25 MPa, the cost of pressurizing from 410 kPa to 25 MPa will consume about 0.3 kWh/Nm^3, or 3% of the power consumed in the upgrading unit.

(D) Marketing and Sales

Like any business, the biomethane product needs to be marketed, sold, invoiced, have quality assurance procedures and customer relations. All of these are costs for the plant owner.

Table 3.4 Compressed biomethane power requirements

Upgrading system technology	Pressure from upgrading process (gauge) (kPa)	Compressor pressure (kPa)	Electricity consumption (kWh/Nm^3)	Compression cost at $0.13/kWh ($/GJ)
Amine wash (COOAB)	15	3300	0.24	0.87
PSA	400	3300	0.12	0.42
Water scrubber	1000	3300	0.063	0.22

3.2.5 *Technology Comparison*

Absorption by water has been the most popular and widespread technology for increasing biogas quality. In 2015, 40% of the upgrading plants were water absorption, see Fig. 3.19. Pressure swing adsorption, membrane separation, and chemical absorption have the remaining share with a small portion (0.4%) going to cryogenic plants. The reason is that water scrubbing is relatively simple and it does not require a pre-cleaning system. European countries including the Czech Republic, Sweden, Switzerland, and the Netherlands mainly use water scrubbing for reducing carbon dioxide and hydrogen sulfide.

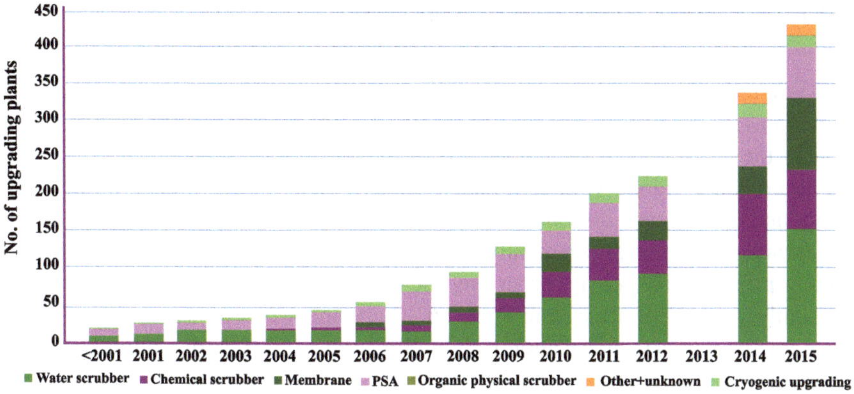

Fig. 3.19 Growth in biogas upgrading units distributed by technology employed (Reprinted with permission from [28] (data from 2013 was not available))

Table 3.5 Advantages and disadvantages of technologies to extract carbon dioxide

System	PSA	Water wash	MEA	Membrane
Absorption characteristics	Absorption	Physical absorption	Chemical absorption	Membrane separation
Pressure (kPa)	400–700	400–700	Atmospheric	1600–4000
Methane loss (%)	3–10	1–2	<0.1	3–15
Pretreatment system	Needed	Not needed	Needed	Needed
End methane concentration (%)	>96	>97	>99	90–94
Energy use (kWh/$Nm^3_{raw\,biogas}$)	0.25	<0.25	<0.15	0.2–0.38
System temperature (°C)	Room	Room	100 °C	Room
Energy used in recovery	Moderate	Moderate	Very high	None
Chemical recovery	Yes	Yes	Yes	–

The final decision on what type of upgrading system to use will depend on the quality of raw biogas entering the system, especially the level of impurities such as H_2S or siloxanes. It will depend on the percentage of pure methane required in the biomethane. Other factors include the availability of fresh water, the price of electricity, local standards and emissions protocols, the production rate of biogas, and tolerance for methane losses (slip) in the process. There is no clear winner with various processes having unique advantages as summarized in Table 3.5.

3.3 Biomethane Post Treatment

3.3.1 Moisture Reduction in Biomethane

Moisture reduction for biogas pretreatment systems has been discussed in Sect. 2.6. Sometimes a cooling system is also desirable post treatment. Having a high humidity reduces the heating value of biomethane. If the moisture in biomethane condenses, the liquid can combine with gases such as CO_2, NH_3, or H_2S, resulting in acidic compounds. They can corrode equipment which uses steel such as piping networks, valves, and gas burners. Its generally considered good practice to reduce the moisture content prior to transporting or storing the biomethane. Some of these techniques are similar to those discussed in Sect. 2.6. There are two common drying methods for biomethane: (1) Condensation Dehumidification, and (2) Adsorption Dehumidification.

3.3.1.1 Condensation Dehumidification

This is where the biogas is cooled below its dew point temperature. The moisture condenses out. The lower the temperature the more condensation. There are two technologies used, a vapor compression refrigeration system or a cold water chiller.

Refrigeration cooling system The system is commonly used in refrigeration and air conditioning. As shown in Fig. 3.20 this cooling system has been described in Sect. 2.6.2.

Chiller system A chiller unit operates the same as a refrigeration cycle. There is an additional heat exchanger that circulates water through the evaporator unit. The chilled water is sent where cooling is required. They are used in large air conditioning systems, such as shopping malls and convention centers. Figure 3.21 shows the layout of a chiller.

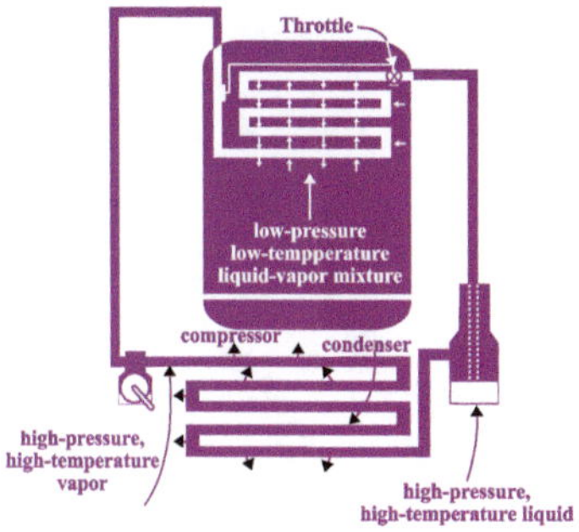

(a) Schematic of a refrigeration unit

(b) Actual refrigeration unit

Fig. 3.20 Vapor compression refrigeration system

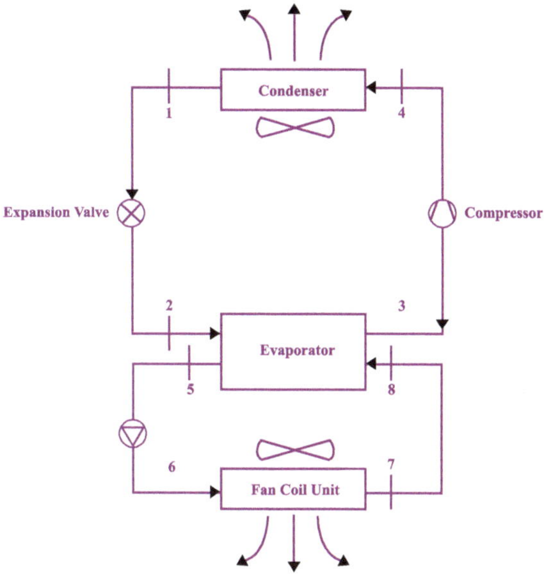

(a) Chiller outline

(b) An actual chiller

Fig. 3.21 Refrigeration chiller unit

3.3.1.2 Adsorption Dehumidification

This process uses a desiccant which is a humidity-absorbing material. Desiccants absorb moisture from the air directly. Once the desiccant is saturated it can be removed and dehumidified, typically by heating it at temperatures between 55 and 150 °C. It can be mounted on a conveyor belt and cycled during operation. They are especially suited for high humidity levels at low temperatures. There is no compressor so they are cheaper, quieter, and lighter than refrigerators or chillers. The examples

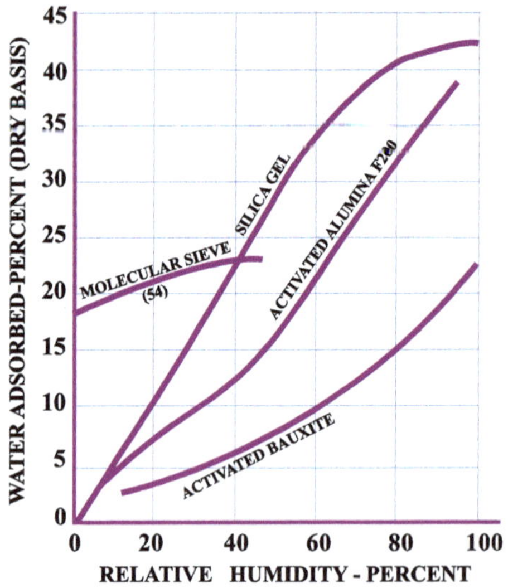

Fig. 3.22 Solid desiccant properties, at atmospheric pressure and a 25 °C temperature. Adapted with permission from [33]

of solid desiccants are silica gels, activated alumina, or molecular sieves. They are hygroscopic materials that have the ability to absorb moisture, as shown in Fig. 3.22.

The equilibrium capacity is the ratio between the mass of moisture adsorbed and the mass of adsorbent is the key parameter in the adsorption process. The equilibrium capacity depends on the temperature and relative humidity. When designing a dehumidification system, equilibrium data should be obtained directly from the manufacturer of the adsorbent used.

Generally, a continuous system is operated by a two-column system. One column removes moisture from the biomethane while the other is regenerated. The adsorption column works for 8–24 h and then it is disconnected from the gas stream and heated to the regenerating temperature. For large-scale plants (<500 Nm^3/h), the regenerating gas is often taken from the slipstream of the dry biomethane gas. After passing through the column, it can is reused by cooling it down in a chiller and removing the condensed water from the biomethane, see Fig. 3.24. For smaller systems, a vacuum pump is used to dry out the saturated column.

The biomethane flow direction is important. It usually flows downward in order to avoid fluidization of the desiccant bed. The upper part of the column is saturated with water from the wet biomethane first. When the amount of water adsorbed on the desiccant is in equilibrium with its partial pressure the zone is called the saturation or depleted zone. Below the saturation zone, there is a section in which mass transfer takes place as water vapor molecules migrate into the adsorbent pores. This zone is called a Mass Transfer Zone (MTZ) and its height is essential for determining the suitable column height. Any biomethane leaving this zone usually has very low moisture content. The last part in the lower section of the column is where the adsorbent in it is still virgin. As the operation continues, the adsorbent in the upper part is depleted, resulted in the MTZ migrating downward. Once the edge of MTZ

Table 3.6 Composition of silica gel

Material	Volume (%)
Silica (SiO_2)	97–99.9
Iron, Fe_2O_3	0.01–0.03
Aluminum, Al_2O_3	0.04–0.1
Sodium, Na_2O	0.02
Titanium, TiO_2	0.03–0.09

reaches the bottom of the column, the moisture in the biomethane product increases. At this point, the operation should cease and be switched to regeneration mode. The regeneration gas flows in the opposite direction in order to drive water from the bottom to the top of the column. So the bottom part of the adsorbent bed is dried first, causing little effect on the product gas quality when the operation resumes. After regeneration, the column is cooled by passing low-temperature gas through it.

Solid desiccant absorption—Silica Gel Silica gel is manufactured from sodium silicate. It is derived from sand with a high percentage of silica, mixed with dilute sulfuric acid, which has moisture absorbing properties. The porous structure of silica causes water retention. Silica gel is clear and glassy and is shaped like quartz. It is sold as small round beads and in powder form. The manufacturing process begins with silica which has a chemical formula, $SiO_2.nH_2O$. It is then reacted with sodium silicate and sulfuric acid. The sodium is removed and after the silica gel is dried. Its composition is shown in Table 3.6. Features of the silica gel are as follows:

- It comes as a clear or a blue pill, is not water-soluble and has a boiling point of 1,900 °C.
- Its size ranges from 2 to 5 mm, with an average pore diameter of between 2 and 4 nm.
- At a temperature of 30 °C and relative humidity 20–90%, silica can absorb 8–30% of its mass in moisture.
- A surface area of 650–800 m^2/g and a density of 0.70–0.78 kg/m^3 at a humidity of $<3\%$.
- Is semipermanent, if stored in the absence of air or moisture and is nonflammable.
- It can be packaged in various sizes, is easy to use, has no taste or smell.
- Can be recycled by heating at a temperature of between 150 and 200 °C for 2 h.
- Is chemically inert.

The balance between moisture in the gas stream and the moisture absorbed by the silica gel is shown in Fig. 3.23. Equation 3.14 can be used to calculate the moisture absorbed by the silica gel. Figure 3.24 is a process diagram of using silica gel for moisture reduction.

$$W = KC^n \tag{3.14}$$

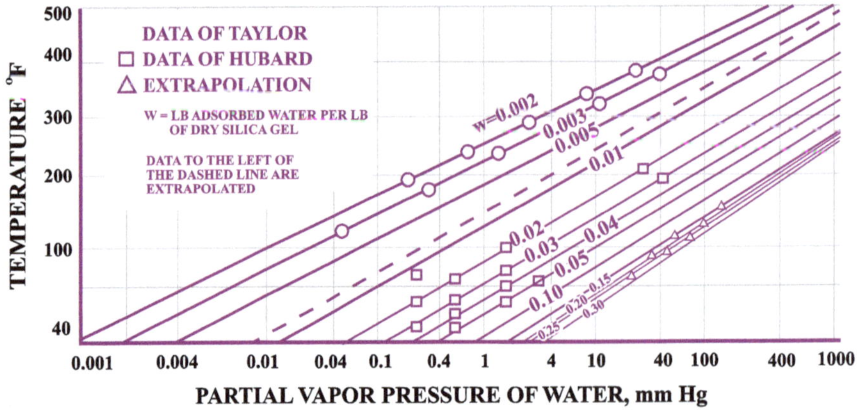

Fig. 3.23 The partial pressure of water vapor in equilibrium with the silica gel, absorption of water at various levels (Adapted with permission from [33])

where

W Concentration of water in silica gel, mol/m^3
C Concentration of water in the biomethane gas, mol/m^3
K Constant, [−]
n Constant, [−]

Silica gel is used in various industries for moisture removal, such as gas, food, pharmaceutical, and electronics. Its equilibrium properties with water vapor are very well understood, Fig. 3.23. Typically there are two desiccant columns. One is drying the biomethane and the other is getting regenerated as shown in Fig. 3.24.

Molecular Sieve A molecular sieve is a type of synthetic zeolite which is a class of inorganic silicate materials. They have compounds of type $(K_2O.Na_2O).Al_2O_3.2SiO_2.xH_2O$. Figure 3.25 shows a molecular sieve, in 2 mm diameter cylindrical shapes with a density between 0.64 and 0.66 kg/m^3. It is tasteless, odorless, nontoxic, and inflammable. It can absorb moisture in cold conditions and desorb it under hot conditions. Under low ambient humidity (10–30% RH) molecular sieves can effectively absorb 22% of their dry weight. They have a surface area of about $700-800\,m^2/g$. They can be regenerated by heating at a temperature of $200-250\,°C$ for 2 h which is relatively high. They do not react with any other gas and can easily be separated. Dehumidification by molecular sieves can provide the lowest dew point compared to alumina and silica gel, see Fig. 3.22. The equilibrium moisture adsorption depends on temperature and relative humidity (%RH). The cost of molecular sieves is usually higher than other types of adsorbents (Fig. 3.26).

Liquid desiccant absorption—Glycol Another method of moisture removal is to use a desiccant liquid such as ethylene glycol (Diethylene Glycol; DEG) or triethylene glycol (Triethylene Glycol; TEG).

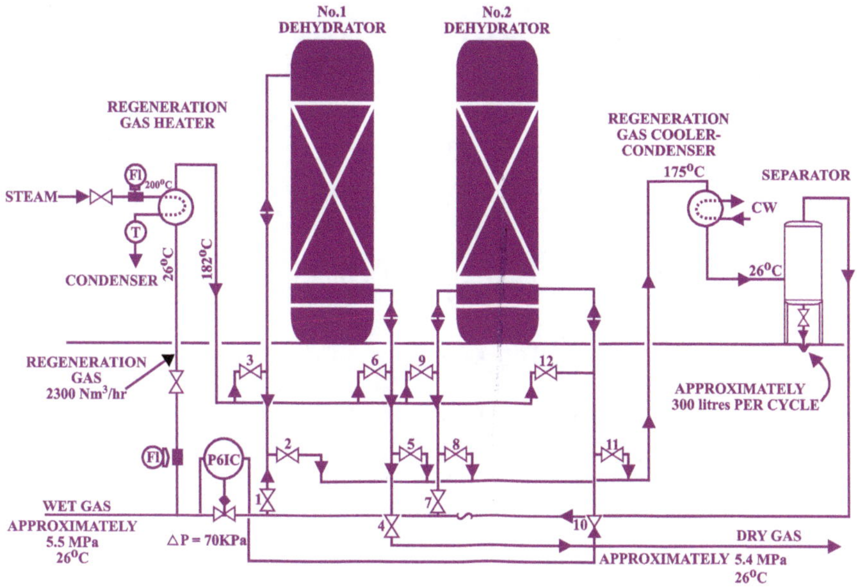

Fig. 3.24 Silica gel use in the gas dehumidification process (Adapted with permission from [33])

A research survey on dehumidification by desiccant liquid was conducted by [34]. The absorber used to absorb moisture in natural gas was triethylene glycol. Modeling and the relationship between the gas flow rate and the purity of the triethylene solution were used to determine the optimal absorber tank size.

3.3.1.3 Desiccant Design Procedure

(1) The column diameter should be determined by selecting a superficial velocity which does not produce too high pressure drop. The maximum pressure drop should not exceed 7.5 kPa/m and the total pressure drop should be less than 35–55 kPa to avoid high energy consumption and adsorbent fracture. The pressure drop can be calculated by using the following modified Ergun's equation, Eq.3.15. Once the allowable superficial velocity is determined, the column diameter can be obtained (Table 3.7).

$$\frac{\Delta P}{L} = B\mu u_{max} + C\rho u_{max}^2 \tag{3.15}$$

where

$\frac{\Delta P}{L}$ Pressure drop per unit length, psi/ft

μ Biomethane viscosity, centipoise

Fig. 3.25 Molecular sieves

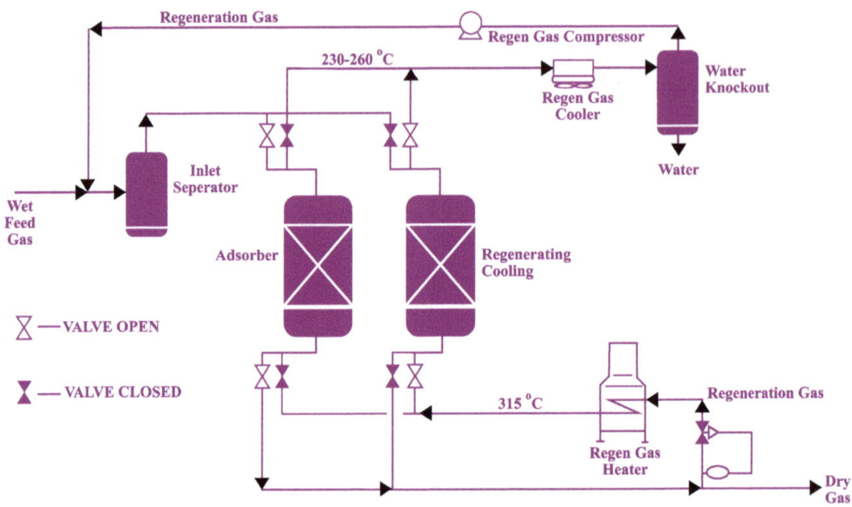

Fig. 3.26 Process diagram for moisture reduction in natural gas, with molecular sieve (Adapted with permission from [33])

Table 3.7 Constants for Ergun's equation

Desiccant particle type	B	C
1/8" Bead (4 × 8 mesh)	0.056	0.0000889
1/8" Extrudate	0.0722	0.000124
1/16" Bead (8 × 12 mesh)	0.152	0.000136
1/16" Extrudate	0.238	0.00021

ρ Biomethane density, lb/ft^3
u_{max} Max. superficial velocity, ft/min
B Constant, min^2/ft^2
C Constant, min^2/ft^2

(2) The mass of adsorbent can be calculated based on the operation period, which is normally between 8 and 12 h before regeneration is necessary. If a longer operation period is selected, a larger column volume requiring more capital investment is needed. The total column height is the summation of the saturation zone height and the MTZ height. The saturation zone height is calculated by the following:

$$\mathit{The\ saturation\ zone\ height}\ (\mathrm{m}) = \frac{\mathit{Total\ adsorbent\ mass}\ (\mathrm{kg})}{\mathit{Column\ cross-sectional\ area} * \mathit{Bulk\ density}\ (\mathrm{kg/m})} \tag{3.16}$$

$$\mathit{The\ adsorbent\ mass}\ (\mathrm{kg}) = \frac{\mathit{Water\ mass\ removed}\ (\mathrm{kg})}{\mathit{Water\ adsorption\ capacity} * \mathit{Correction\ factor}\ (-)} \tag{3.17}$$

The height of the MTZ can be calculated using the following equation:

$$L_{MTZ} = \left(\frac{Gas\ velocity}{35}\right)^{0.3} Z \tag{3.18}$$

where Z: = 0.170 ft for 1/8 inch sieve
Z: = 0.85 ft for 1/16 inch sieve

An extra 3 ft (0.9 m) must be added to the column height to allow for space for biomethane inlet and outlet distribution.

(3) The pressure drop is proportional to the square of the superficial velocity. The following equation (Eq. 3.19) can be used to calculate the actual pressure drop. The actual pressure drop occurs in the commercial cylinder based on the actual diameter selected which is normally rounded up from the calculated diameter. The actual adjusted superficial velocity is also calculated from the commercial diameter.

$$\left(\frac{\Delta P}{L}\right)_{adjusted} = (0.33\ psi/ft)\left(\frac{u_{adjusted}}{u_{max}}\right)^2 \tag{3.19}$$

Once the pressure drop per column height ($\Delta P/L$) is selected, the total pressure drop can be obtained by multiplying it with the column height. The design pressure drop should be in the range of 5–8 psi (34.5–55.1 kPa) in order to avoid too high energy consumption and excessive adsorbent abrasion. If the design pressure is too high, a larger column diameter can be selected.

(4) The regeneration can be conducted using a temperature or a vacuum process. In thermal regeneration, the minimum total heat required is the sum of the heat required for the desorption of water (Q_w), the heat for raising the adsorbent temperature to the regeneration temperature (Q_s), the heat for raising the cylinder and piping temperature to the regeneration temperature (Q_{st}) and the heat loss (Q_{hl}) which is approximately 10% of the total heat. The actual heat used should be 250% of the minimum heat calculated to compensate for the effects of temperature change during the regeneration period.

Example Calculation After biogas upgrading by a water scrubbing process, the biomethane still contains some amount of moisture. The flow rate is $300\,Nm^3/h$ at standard conditions. The actual pressure is 600 kPa and temperature 40 °C. The adsorbent selected is a molecular sieve (Type 4A) in a form of 1/16" beads.

Step (1) Using Eq. 3.15 determine the maximum allowable superficial velocity that provides the pressure drop of 0.33 psi/ft. The biomethane viscosity is $\mu =$ 0.012 centipoise and density is $\rho = 0.294\,lb/ft^3$. So:

$$u_{max} = 99.5\,ft/min = 30.33\,m/min$$

The volumetric flow rate of is $300\,Nm^3/h$ at normal conditions. At the actual conditions (T = 40 °C and P = 600 kPa), the volumetric flow rate is

$$\dot{V}_{actual} = \dot{V}\left(\frac{P_{std}}{P_1}\right)\left(\frac{T_1}{T_{std}}\right)\left(\frac{z}{z_{std}}\right)$$

$$\dot{V}_{actual} = 300\left(\frac{101.325}{600}\right)\left(\frac{313}{273}\right)\left(\frac{0.93}{1}\right) = 54\,m^3/h$$

The minimum column diameter can be calculated from:

$$D_{min} = \left(\frac{\dot{V}_{actual}}{0.25\pi\, v_{max}}\right)^{0.5} = 0.19\,m$$

Rounding this up to a standard commercial size column gives a diameter of $D_{column} = 0.3\,m$. The adjusted superficial velocity, based on the actual diameter can be calculated from:

$$u_{adjust} = \left(\frac{\dot{V}_{actual}}{0.25\pi D^2}\right) = 0.21\,\text{m/s}\ or\ 41.7\,\text{ft/min}$$

Step (2) is to calculate the amount of water removed, given a 12 h cycle time. The amount of water in the inlet biomethane can be determined based on its dew point temperature. In this case, the dew point of the biomethane leaving the upgrading unit is 10 °C at an STP equivalent to 6.5 g of water in 1 Nm^3 of biomethane. The dew point is needed to be reduced to −50 °C, equivalent to 0.04 g/Nm^3. Therefore, the amount of water required to be removed is

$$Water\ to\ be\ removed = 300\,\text{Nm}^3/\text{h} \times (6.5 - 0.04)\text{g/m}^3 \times 1/1000\,\text{kg/g} = 1.94\,\text{kg/h}$$

So the water removed per cycle (12 h) is $Water\ removed = 23.26\,\text{kg/cycle}$.

Step (3) Find a length of the mass transfer zone (LMTZ)

$$L_{MTZ} = \left(\frac{Gas\ velocity\ (\text{ft/min})}{35}\right)^{0.3} (Z) = 1.34\,\text{ft}$$

This is a very short LMTZ and usually minimum length of 6 ft (1.82 m) is chosen.

Step (4) is to calculate the amount of adsorbent required. Providing that a kg of molecular sieve can adsorb moisture up to 20% of its total weight.

$$Adsorbent\ required = \frac{Water\ removed}{Adsorbent\ capacity} = \frac{23.26}{0.2} = 116.3\,\text{kg}$$

The bulk density of the molecular sieve is $\rho_{ms} = 720\,\text{kg/m}^3$. The bed height (Lms) is then:

$$L_{ms} = \frac{4 \times Adsorbent\ weight}{Bulk\ density\ \times \pi D^2} = \frac{4(116.3)}{720\pi 0.3^3} = 2.28\,\text{m}$$

There, the total bed height is $L_{total} = L_s + L_{MTZ} = 2.28 + 1.82 = 4.1\,\text{m}$. Roundup Ltotal to 4.50 m. This total column height must be added with an extra 1 m space for column internals, bringing Ltotal to 5.50 m.

Step (5) Is to check the total pressure drop through the column. Equation 3.19 can be used to calculate the pressure drop:

$$\left(\frac{\Delta P}{L}\right)_{adjusted} = (0.33\,\text{psi/ft})\left(\frac{u_{adjusted}}{u_{max}}\right)^2 = 0.33\left(\frac{42.10}{99.50}\right)^2 = 0.059\,\text{psi/ft}$$

Along the bed length (5.5m or 18ft), the total pressure drop is $= 0.059 \times 18 = 1.06$ psi or 7.3 kPa. This is less than the requirement of 34.5 kPa and so these parameters and dimensions are acceptable in the desiccant column design.

3.3.2 Odorization

Adding an odor to biomethane is necessary in order for it to be easily detected in case it leaks. Gas odorization is a safety-critical issue in natural gas distribution industries. Methane is colorless and odorless. For safety purposes, it is necessary to add an odorizer in order to be able to smell the gas. Odorizers typically used include tetrahydrotiophen or ethylene mercaptan (ethanethiol). Usually, it has to be scented by an average person in quantities that are one fifth below the flammability limit, according to the International Organization for Standardization number ISO

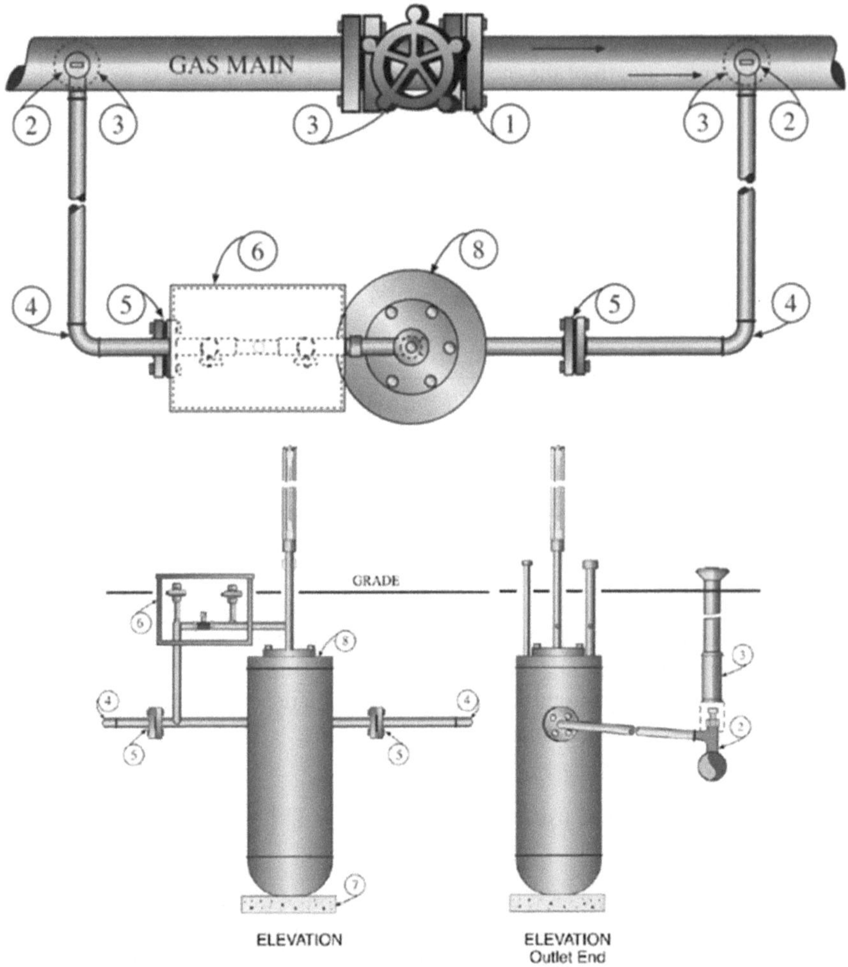

Fig. 3.27 General installation features for natural gas odorizers (with permission from King Tool Co., http://www.kingtoolcompany.com/wp-content/uploads/2012/11/Gas-Odorizers.pdf)

Table 3.8 Device details for Fig. 3.27 (King Tool Co.)

Item number	Quantity	Material
1	1	Orifice plate
2	2	Angle gate valve
3	3	Valve handle
4	2	90° Elbows
5	2	Flanges
6	1	Valve box
7	1	Concrete box
8	1	King tool odorizer

13734 [35]. The odorant must not cause harm to people, materials or the pipe material. The products of combustion from the odorant should not be toxic when breathed nor should they be corrosive or harmful to materials to which the products of combustion will be exposed. The odorant should not be soluble in water to an extent greater than 2.5 parts to 100 parts by weight in case moisture in the pipeline diminishes the odorant smell. The operator should carry out periodic "sniff" tests at the ends of the system to confirm that the gas contains odorant. In large natural gas pipes, odorizers are connected in parallel with the pipeline. A small quantity of natural gas is bypassed through the odorizer. They are designed to allow the small amount of gas to absorb enough odorant to produce the desired odor intensity in the main gas line. The odorizer shown in Fig. 3.27 is from King Tool and is to be used only for pipeline quality biomethane in a continuous flow process (Table 3.8).

3.4 Case Study—Water Scrubbing Plant

A pilot-scale biogas upgrading system based on pressurized water scrubbing technology was installed and commissioned in a chicken farm in Chiang Mai, Thailand, to convert ordinary biogas from chicken manure to biomethane at a production capacity of 20 Nm^3/h of biomethane, at an absorber pressure of 400 kPa (gauge). The compressed biomethane production system consists of the following basic components, see Fig. 3.28

- Biogas digester,
- Biogas cleaning system,
- Biogas upgrading system (water scrubber),
- CNG compressor and storage facility, and
- Dispenser.

This system is described by [36]. The water scrubber comprises two main parts: an absorber unit and a desorber unit which includes a flash tank and a desorber, as

Fig. 3.28 Biomethane production plant

shown in Fig. 3.29. The 40 cm absorption tank made from stainless steel was packed with 50-mm pall rings of 2 m height. A two-piston type gas compressor (SWAN; SVP-203) was used with a 2.2 kW explosion-proof motor (Crompton). The flash tank and the desorber were stainless steel vessels with the diameters of 60 cm and 30 cm, respectively. The desorber was internally packed also with 50-mm pall rings with 1.5 m packing thickness. After the upgrading process, the upgraded biogas was then dehumidified using a shell and tube vapor condenser and a silica gel adsorption unit.

The system is able to run continuously with good stability and produce biomethane containing approximately 85% methane with a Wobbe Index of 40.35 MJ/m^3. The biomethane is then compressed to 20 MPa into a cylindrical fuel vessel for use in domestic or transportation applications.

3.4.1 Design Assumptions

1. The biogas input to the upgrading unit is assumed to have the properties of typical biogas derived from the pretreatment unit of chicken farm as can be seen in Table 3.9.

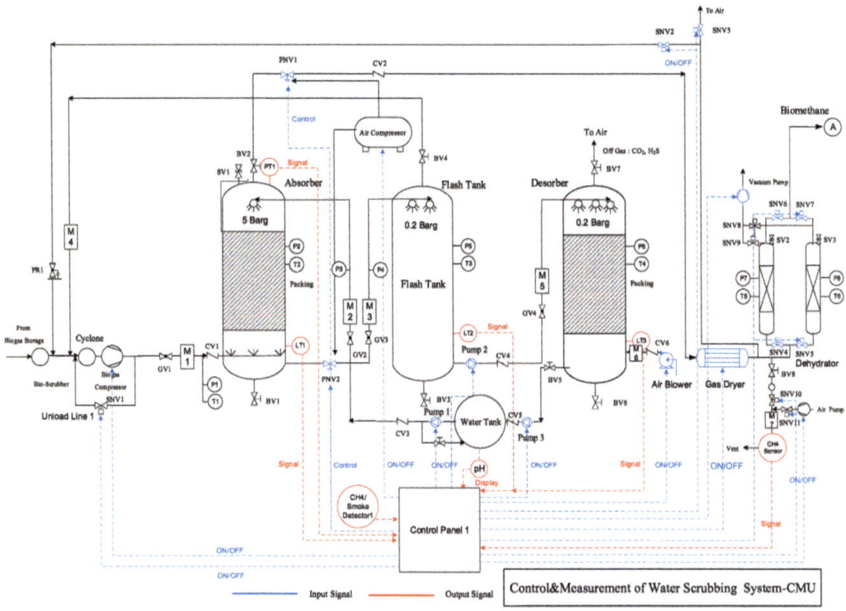

Fig. 3.29 Schematic diagram of a pressurized water biogas scrubbing system [36]

Table 3.9 Biogas properties from swine farms

Element	Concentration
Methane	50–65% by Vol.
Carbon dioxide	25–45% by Vol.
Oxygen	<0.3% by Vol.
Hydrogen sulfide	<200 ppm
Hydrogen	0.5–1.0% by Vol.
Moisture	80–100% RH at 40 °C
Biogas flow rate	40 Nm^3/h

2. The inlet pressure of the biogas should not be greater than the maximum allowable inlet pressure to the compressor. The output pressure is going to be equal to the pressure in the absorber minus the piping pressure drop, which is in the range of 300–500 kPa.
3. The properties of biomethane produced from the plant are shown in Table 3.10.
4. The water used in the plant must be capable of preventing the growth of microorganisms. It is necessary to control the hardness and amount of contaminants in the water. The water properties are shown in Table 3.11.

Table 3.10 Biomethane properties from the absorber exit

Element	Concentration
Methane	85–90% by Vol.
Carbon dioxide	7–14% by Vol.
Oxygen	<1.0% by Vol.
Hydrogen sulfide	0–10 ppm
Hydrogen	<0.7% by Vol.
Biomethane flow rate	20 Nm^3/h

Table 3.11 Water properties in the water scrubber

Element	Concentration
Residual chlorine	0.2–0.5 mg/l
pH	7.0–8.0
Total alkalinity	<300 mg/l as $CaCO_3$
Total dissolved solids	<600 mg/l

3.4.2 *General System Considerations*

There are some basic parameters for designing industrial tanks and piping. The following standards should be following and used as tools to aid the system design:

- The design of pressure tanks should be based on ASME Code VIII Div. 1-boiler and pressure vessel code [37].
- The design of gas and liquid pipes should be based on the ANSI/ASME B31.3 [38] process piping design standards. The flow through valves and joints in the system can be modeled as single-phase flow using the Darcy–Weisbach equation (Eq. 3.20) for preliminary calculations

$$\frac{\Delta P}{L} = f_D.\frac{\rho}{2}.\frac{v^2}{D} \tag{3.20}$$

- where the pressure loss per unit length $\Delta P/L$ (Pa/m) is a function of

ρ, The density of the fluid (kg/m^3)
D, The hydraulic diameter of the pipe (m)
v, The mean flow velocity, (m/s)
f_D, The Darcy friction factor

- The selection of the pump system can be based on the Net Positive Suction Head (NPSH) criteria and the performance curve of the pump. The appropriate standards are the ANSI/AWWA E103-07 [39]. Alternatively, the API Standard 610 [40] for centrifugal pump may also be used. All pumps must be explosion proof.
- The compressors should be selected based on the API Standard 618, Reciprocating Compressors for Petroleum, Chemical, and the Gas Industry [41].

- There are 20 gas cylinders for biomethane storage. They are designed based on ISO 11439 or [37] (Boiler and Pressure Vessel Code). The high-pressure compressor and fueling connection devices used should follow the ANSI/NGV 1 standard [42].

3.4.3 Water Scrubbing Design

The operation of a biomethane upgrading system by means of water scrubbing is shown in Fig. 3.29. Starting from the pump, the water is piped to the spray head in the absorption tank. The nozzle head sprays water over the absorption cross-sectional area and exits at the bottom.

At the same time, the compressor drives the pretreated biogas from the gas storage tank into the absorber. It is distributed at the bottom of the absorber through a distributor. The biogas pressure in the absorption tank is maintained at 400 kPa. The hydrogen sulfide gas is treated to below 200 ppm in the bioscrubber. The remaining hydrogen sulfide gas and carbon dioxide in the biogas are absorbed by water in the absorption tank, along with some methane. The biomethane exits through the top of the absorption tank. The water, exiting the bottom of the tank, contains increased levels of hydrogen sulfide and carbon dioxide. The relationship between the water flow, tank pressure, and biomethane purity was simulated for a biogas flow rate of 40 Nm^3/h and shown in Fig. 3.30. A purity level of 90% was required, so from Fig. 3.30 a water flow rate of 40 m^3/h and an absorber pressure of 400 kPa was selected.

The water is depressurized in a flash tank to 100 kPa_{gauge}. This allows most of the dissolved methane to be recovered. This gas is recycled to a port before the compressor and repeats the cycle. This step minimizes the methane slip. Figure 3.31 shows a simulation of the pressure in the flash tank versus the final methane content in the biomethane. If the flash tank pressure falls below 1 bar_{gauge} (100 kPa_{gauge}), carbon dioxide begins to be released along with the methane. This gets fed back to the absorber and results in more carbon dioxide in the treated output. If the flash tank pressure is above 1 bar_{gauge} (100 kPa_{gauge}), this increases the methane slip as less methane escapes. The optimal choice of flash tank pressure is therefore 100 kPa_{gauge}.

The next stage for the CO_2-rich water is the desorber. The pressure is reduced further to 10 kPa_{gauge}. This removes most of the hydrogen sulfide and carbon dioxide gases. The desorber is equipped with an external air blower to help extract the dissolved gases. The gases enter the atmosphere. The CO_2 lean water falls to the bottom of the desorber. It flows to a tank where it is stored until ready to begin a new flow loop and the process repeats. The water absorption design is based on the GPSA Engineering Data Book of the Gas Processors Suppliers Association [43]. The design outline is as follows.

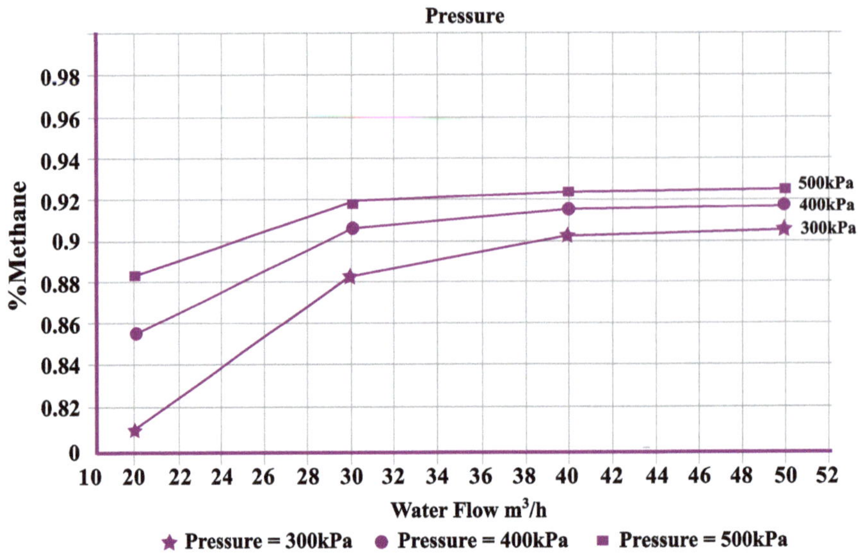

Fig. 3.30 Water flow rate and pressure versus biomethane purity

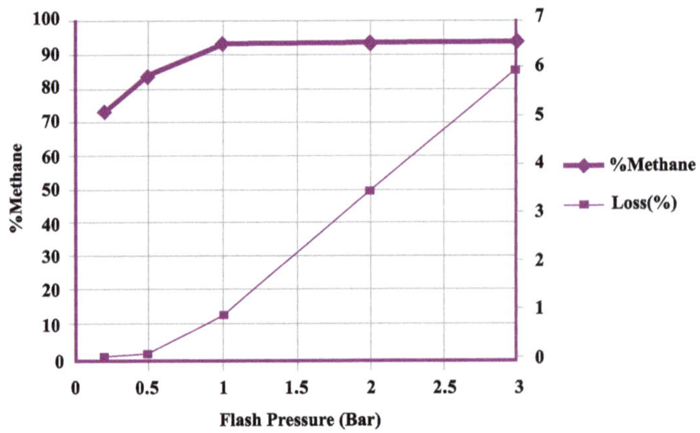

Fig. 3.31 Flash tank pressure versus final methane content

1. Use the vapor liquid equilibrium data (VLE) between carbon dioxide, hydrogen sulfide, and water. Based on experimental data from [44], or simulated results with Aspen HYSYS, ChemCAD, or equivalent software.
2. The design of the absorber and the desorber uses the McCabe–Thiele method [45]. The absorption of the gases is a physical absorption method, with a countercurrent flow regime between the biogas and water. This method gives the minimum water flow rate required to remove a certain percentage of carbon dioxide.

Table 3.12 Final specifications for the absorption tank

Description	Specification
Working pressure	400 kPa
Design pressure	800 kPa
Design temperature	40 °C
Max. allowable stress	116,000 kPa
Internal diameter	35 cm
Packing height	155 cm
Total height	390 cm
Total volume	0.375 m^3
Thickness	4.5 mm
Material	Carbon steel A516 grade 55
Finishing	Coated with epoxy vinyl ester

3. The absorption tank should be designed to accommodate a water flow rate of 1.5 times the minimum flow rate determined from McCabe–Thiele. This is the optimal flow rate for biogas purification [46]. The absorber final specifications are shown in Table 3.12.
4. The absorber sizing should be designed using the Sherwood-Eckert method [47] with flooding of up to 70% of the system flow rate and a pressure drop of no more than 1225 Pa/m. These are standard rules of thumb from designers familiar with the field.
5. The flash drum is designed based on a vertical separator, with constant pressure and a maximum upflow gas velocity. Typical vertical separators have a length to diameter ratio in the 2–5 range. Exact design details can be found in [43]. The flash drum final specifications are shown in Table 3.13.
6. The desorber design is again based on the McCabe–Thiele method. The carbon dioxide in the water escapes and passes through a gas spillway, leaving no more than 0.2% of the initial concentration in the remaining water. The desorber final specifications are shown in Table 3.14. The actual physical absorber, desorber, and flash tank are shown in Fig. 3.32.

The exit biomethane gas from this water scrubbing system contains high level of water vapor, since the gas has just passed through water. It becomes necessary to remove the moisture by two steps.

1. A chiller is used to reduce the temperature of biomethane. The theoretical details have been provided in Sect. 2.6.2. At or below the dew point temperature the vapor will condense. In this case study, the chiller is a shell and tube design made from stainless steel and aluminum. The design can support a flow rate of 63 Nm^3/h with an incoming biomethane temperature of 35 °C, 100% RH and a 2 °C output temperature. The refrigerant used is R-134a and the system is controlled by a

Table 3.13 Final specifications for the flash drum

Description	Specification
Working pressure	100 kPa
Design pressure	400 kPa
Design temperature	40 °C
Max. allowable stress	1,160 kg/cm^2
Internal diameter	60 cm
Total height	290 cm
Total volume	0.82 m^3
Thickness	4.0 mm
Material	Carbon steel A516 grade 55
Finishing	Coated with epoxy vinyl ester

Table 3.14 Final specifications for the desorption tank

Description	Specification
Working pressure	10 kPa
Design pressure	400 kPa
Design temperature	40 °C
Max. allowable stress	1,160 kg/cm^2
Internal diameter	60 cm
Total height	290 cm
Total volume	0.82 m^3
Thickness	3 mm
Material	Carbon steel A516 grade 55
Finishing	Coated with epoxy vinyl ester

thermostat electronic control. The water pump sends the cold water to the shell & tube coil unit with high humidity biogas flowing through it (Fig. 3.33).

2. Since the biomethane moisture content is so high a second drying system is employed. After the chiller, a dehydrator further reduces the moisture content of the biomethane. The dehydrator design is cylindrical with an inner diameter of 25 cm high and a packing height of 230 cm. It is a pressure vessel with a pressure drop of not more than 34.5 kPa. The working temperature is less than 10 °C. The cylinder material is carbon steel A516. For safety, two pressure relief valves designed specifically for biomethane are used on each cylinder. They open at a pressure of 500 kPa and are designed for a maximum flow rate of 20 Nm^3/h (Table 3.15 and Fig. 3.34).

(a) Absorber

(b) Desorber

(c) Flash Tank

Fig. 3.32 Components of the biomethane water scrubbing system

(a) Condenser

(b) Evaporator

Fig. 3.33 Components of the biomethane post treatment chilling system

Table 3.15 Dehydrator specifications

Description	Specification
Working pressure	400 kPa
Design pressure	800 kPa
Max. temperature	40 °C
Max. allowable stress	116, 000 kPa
Internal diameter	25 cm
Packing height	230 cm
Total volume	0.11 m^3
Quantity	Two tanks
Material	Carbon steel A516 grade 55 coating epoxy vinyl ester+E glass#300
Standard	ASME code VIIII Div.1

3.4.4 Civil Engineering Design

The building has a reinforced concrete base and the steel structure is mounted on this base. All building materials must be fire resistant and nonhazardous if combusted. The design guidelines are as follows:

- The design of the reinforced concrete foundation used the American Concrete Institute (ACI) 350-01 [48] Specifications for Environmental Engineering Concrete Containment Structures. Corrosion-resistant concrete is used.
- The wall between the production of biomethane and the high-pressure biomethane storage must be a fire wall.
- The building uses soundproof insulation, which helps reduce noise levels from the building.
- The building is higher than ground level to protect from rain and flooding. The floor is at a slight angle in case water is spilled inside. There is a drainage system inside with a sump pump in case of accidental flooding (Fig. 3.35).

3.4.5 Electrical System Design

The electrical system design includes the power system design for the water pump, blower motor, and compressors. There will be a Main Distribution Board (MDB) cabinet which is the power control center of the system, see Fig. 3.36. A Programmable Logic Controller (PLC) is used for control. The power cords used throughout the design must be sized correctly for the maximum current. Careful routing of the power lines is needed, a suitable distance away from any potential flammable sources. Any

Fig. 3.34 Desiccant cylinders for biomethane drying

lighting or service electric sockets should be industrial grade, fire, and explosion proof.

The PLC has four major functions:

1. Absorber—Automatic water level adjustment to keep the water level constant. The water level is sensed with liquid capacitance sensors and an automatic valve is adjusted to supply and turn off the water, see Fig. 3.37b.
2. Flash drum—The water level is automatically adjusted to maintain the water level at a constant height in the same way as in the absorber.

Fig. 3.35 Plant civil engineering design

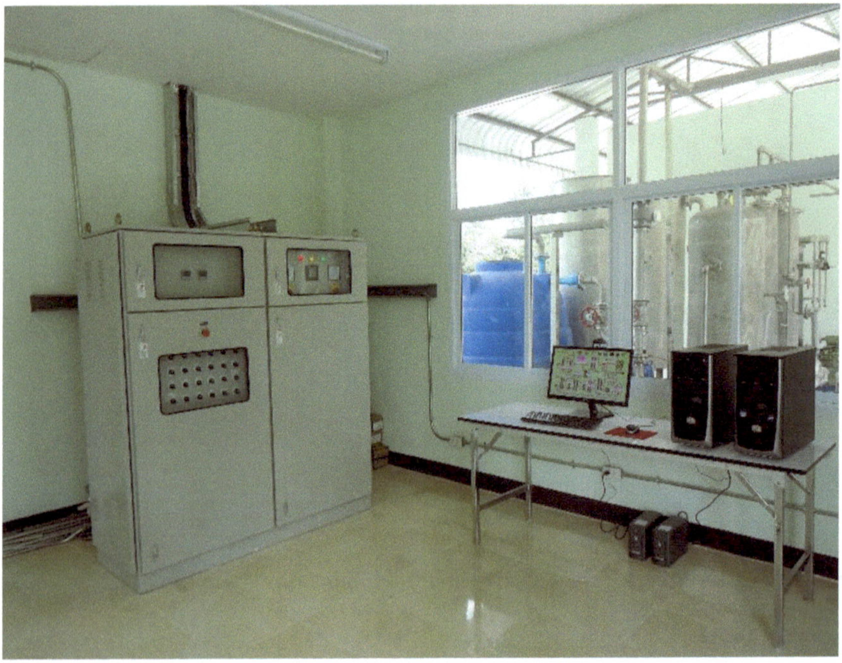

Fig. 3.36 Electrical control room for plant

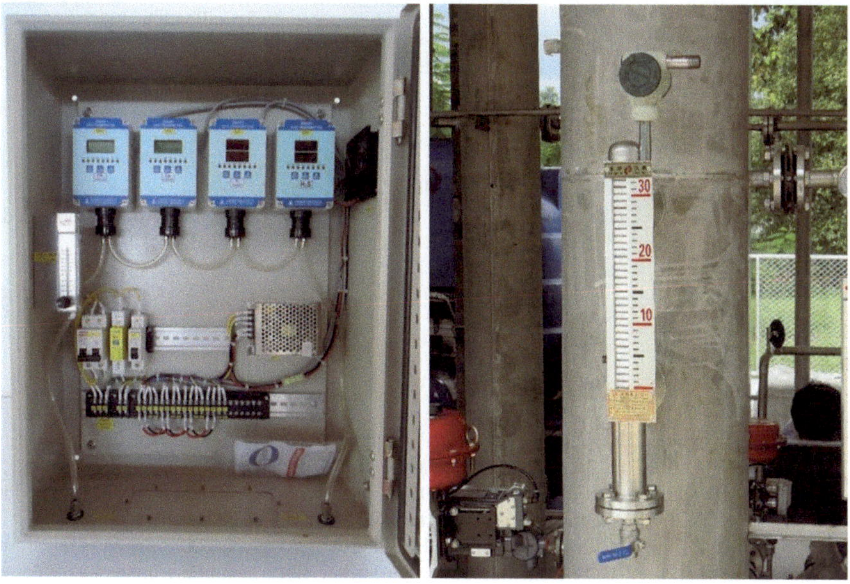

(a) Gas analyzer for biomethane purity inspection (b) Liquid level sensor on the absorber

Fig. 3.37 Gas analyzer and absorber liquid level sensor

Table 3.16 Specifications for the gas analyzer

Description	Specification
Gas/range	$CH_4$0-100% by Vol. $CO_2$0–100% by Vol. $O_2$0–25% by Vol. H_2S 0–5000 ppmv
Measurement technique	CH_4- non-dispersive infrared (NDIR) CO_2- (NDIR) O_2– electrochemical (ECD) H_2S – (ECD)
Accuracy	±3% FS for CO_2, CH_4, O_2 and ±0.5% FS for H_2S

3. Stripper—The PLC turns on/off both pump and air blower automatically using both liquid capacitive sensors to control water level.
4. Gas Analyzer—This sensor array can be set to monitor CH_4, H_2S and the moisture level as a condition to open or close the output valve. If the final biomethane quality is sufficient, the PLC opens the path to the high-pressure compressor. This ensures that only biomethane that meets the final specifications can progress to the storage vessels see Fig. 3.37a and Table 3.16 for details.

The fire alarm system should be equipped with heat detectors, smoke detectors, and gas leak detectors. The effect of any alarm on the system should be the immediate shut down of the plant. The gas leak detector details are shown in Table 3.17.

Table 3.17 Specifications for the gas leak detector

Description	Specification
Applicable gas	Methane CH_4
Range	0–100% lower explosion level (LEL)
Alarm point	20–25% LEL
Response time	<20 s
Amount	Three units (one outdoor and two indoor)

References

1. Authur W, Anna L (2006) Biogas upgrading and utilization. Technical report, IEA Bioenergy
2. Krich K, Augenstein D, Batmale JP, Benemann J, Rutledge B, Salour D (2005) Technologies for removal of carbon dioxide in biomethane from dairy waste, a sourcebook for the production and use of renewable natural gas in California. USDA Rural Development
3. Yang H, Xu Z, Fan M, Gupta R, Slimane RB, Bland AE, Wright I (2008) Progress in carbon dioxide separation of CO2 from flue gas using hollow fiber membrane contactors without wetting. J Environ Sci 20:14–27
4. Kapdi SS, Vijay VK, Rajesh SK, Rajendra R (2004) Biogas upgradation and utilization as vehicle fuel. In: The joint international conference on "Sustainable energy and environment (SEE)", Hua Hin, Thailand
5. Onda K, Takeuchi H, Okumoto Y (1968) Mass transfer coefficients between gas and liquid phases in packed columns. J Chem Enginering Jpn
6. Nock William J, Walker Mark, Kapoor Rimika, Heaven Sonia (2014) Modeling the water scrubbing process and energy requirements for CO2 capture to upgrade biogas to biomethane. Ind Eng Chem Res 53:12783–12792
7. Persson M (2003) Evaluation of upgrading techniques for biogas. Technical report report SGC 142, Swedish gas center
8. Kapdi SS, Vijay VK, Rajesh SK, Prasad R (2005) Biogas scrubbing, compression and storage : perspective and prospectus in Indian context. Renew Energy 30:1195–1202
9. Persson M (2007) Biogas upgrading and utilization as vehicle fuel. In: European biogas workshop, Esbjerk, Denmark, June 2007
10. Electrigaz Technologies (2008) Feasibility study - biogas upgrading and grid injection in the fraser valley. Technical report, BC Innovation Council, British Columbia
11. IUPAC - NIST Solubility Database (2019) https://srdata.nist.gov/solubility/sol_sys_lst.aspx?goBack=Y&sysID=62110&SerialID=DS16
12. Rasi S (2009) Biogas composition and upgrading to biomethane. Master's thesis, University of Jyvaskyla, Jyvaskyla Finland
13. Bansal Pradeep, Marshall Nick (2009) Feasibility of hydraulic power recovery from waste energy in biogas scrubbing processes. Appl Energy 87(3):1048–1053
14. Cavenati S, Grande CA, Rodrigues AE (2005) Upgrade of methane from landfill gas by pressure swing adsorption. Energy Fuels 19(6):2545–2555
15. Miltner M, Makaruk A, Harasek M (2017) Review on available biogas upgrading technologies and innovations towards advanced solutions. J Clean Prod. https://doi.org/10.1016/j.jclepro.2017.06.045
16. Skarstrom CW (1960) Method and apparatus for fractionating gaseous mixtures by adsorption
17. Peterssen A, Wellinger A (2009) Biogas upgrading technologies - developments and innovations. Technical report, IEA Bioenergy

18. Reynolds AJ, Verheyen TV, Adeloju SB, Meuleman E, Feron P (2012) Towards commercial scale postcombustion capture of CO2 with monoethanolamine solvent: key considerations for solvent management and environmental impacts. Environ Sci Technol 46:3643–3654
19. Lepaumier H, Picq D, Carrette PL (2009) New amines for CO2 capture. I. Mechanisms of amine degradation in the presence of CO2. Ind Eng Chem Res 48:9061–9067
20. Maceiras R, Alvarez E, Angeles Cancela M (2008) Effect of temperature on carbon dioxide absorption in monoethanolamine solutions. Chem Eng J 138(1–3):295–300. https://doi.org/10.1016/j.cej.2007.05.049
21. Barzagli Francesco, Lai Sarah, Mani Fabrizio, Stoppioni Piero (2014) Novel non-aqueous amine solvents for biogas upgrading. Energy Fuels 28:5252–5258. https://doi.org/10.1021/ef501170d
22. Brettschneider O, Thiele R, Faber R, Thielert H, Woznya G (2004) Experimental investigation and simulation of the chemical absorption in a packed column for the system NH3 - CO2 - H2S - NaOH - H2O. Sep Purif Technol 39(3):139–159. https://doi.org/10.1016/S1383-5866(03)00165-5
23. Bang JH, Jang YN (2013) Method of producing carbonate using carbon dioxide microbubbles and carbonate therefor
24. Tippayawong N, Thanompongchart P (2010) Biogas quality upgrade by simultaneous removal of CO2 and H2S in a packed column reactor. Energy 35:4531–4535. https://doi.org/10.1016/j.energy.2010.04.014
25. Diao Yong-Fa, Zheng Xian-Yu, He Bo-Shu, Chen Chang-He, Xu-Chang Xu (2004) Experimental study on capturing CO2 greenhouse gas by ammonia scrubbing. Energy Convers Manag 45(13–14):2283–2296
26. Kapoor Rimika, Ghosh Pooja, Kumar Madan, Vijay Virendra (2019) Evaluation of biogas upgrading technologies and future perspectives: a review. Environ Sci Pollut Res 26:11631–11661
27. Gomez-Diaz D, Navaza JM, Sanjurjo B, Vazquez-Orgeira L (2006) Carbon dioxide absorption in glucosamine aqueous solutions. Chem Eng J 122:81–86
28. Hoyer K, Hulteberg C, Svensson M, Jernberg J, Norregard O (2016) Biogas upgrading - technical review. Technical report, ENERGIFORSK
29. Sanders DF, Smith ZP, Guo R, Robeson LM, McGrath JE, Paul DR, Freeman BD (2013) Energy-efficient polymeric gas separation membranes for a sustainable future: a review. Polymer 54:4729–4761
30. ERDI (2013) Final report for biogas promotion from livestock and industrial waste. Technical report, Energy Research and Development Institute, Chiang Mai (in Thai)
31. Beil M, Beyrich W (2013) The Biogas Handbook, Chapter 15, pp 342–378. Woodhead Publishing Limited, Sawston. https://doi.org/10.1533/9780857097415.3.342
32. Cucchiella Federica, D'Adamo Idiano, Gastaldi Massimo (2019) An economic analysis of biogas-biomethane chain from animal residues in Italy. J Clean Prod 230:888–897
33. Kohl Arthur L, Nielsen Richard B (1997) Gas Purification, 5th edn. Gulf Professional Publishing, Houston. ISBN 978-0-88415-220-0
34. Bahadori Alireza, Vuthaluru Hari B (2009) Simple methodology for sizing of absorbers for TEG (triethylene glycol) gas dehydration systems. Energy 34:1910–1916
35. Technical Committee (2013) Natural gas - organic components used as odorants - Requirements and test methods
36. ERDI (2014) A study on the potential of biogas from energy crops as a replacement of LPG. Technical report, Energy Research and Development Institute, Chiang Mai (in Thai)
37. ASME (2019) Bpvc section viii-rules for construction of pressure vessels division 1
38. ASME (2017) Process piping
39. American Water Works Association (2008) Horizontal and vertical line-shaft pumps
40. American Petroleum Institute (2010) Centrifugal pumps for petroleum, petrochemical and natural gas industries
41. American Petroleum Institute (2008) Reciprocating compressors for petroleum, chemical and gas industry services

42. American National Standards Institute (2017) Compressed natural gas vehicle (NGV) fueling connection devices
43. Gas Processors Association (2004) GPSA engineering data book. Gas Processors Suppliers Association, 12 edn,6526 E. 60th St. Tulsa, Oklahoma 74145
44. Jones JH, Froning HR, Claytor EE Jr (1954) Solubility of acidic gases in aqueous monoethanolamine. Chem Eng Data 4(1):85–92. https://doi.org/10.1021/je60001a012
45. McCabe WL, Thiele EW (1925) Graphical design of fractionating columns. Ind Eng Chem 17:605–611. https://doi.org/10.1021/ie50186a023
46. Seader JD, Henley EJ, Keith D (2010) Separation process principles, 3 edn. Wiley, New York. ISBN 978-0470481837
47. Sherwood TK, Shipley GH, Holloway FL (1938) Flooding velocities in packed columns. Ind Eng Chem 30(7):765–769
48. American Concrete Institute (2001) Code requirements for environmental engineering concrete structures

Chapter 4
Biomethane in Transportation Applications

4.1 Introduction

As previously discussed, biomethane is a renewable fuel produced by upgrading raw biogas. Depending on the process, the output quality can satisfy any specification for natural gas quality. For transportation applications, biomethane can be produced to meet all technical requirements set by vehicle manufacturers and natural gas transportation system operators.

Biomethane can be adapted to suit the particularities of very different vehicles. Compressed biomethane is a safe, flexible, and widely available renewable fuel for private cars, transport buses, and heavy-duty vehicles. Britain, France, and Spain plan to ban the sale of petrol and diesel cars from 2040 with Denmark banning the sale from 2030. Biomethane, therefore, offers a solution to reduce gasoline and diesel use. However, infrastructure limitations obstruct implementation. As an EU Directive 2014/94 defined it: "[the lack of] alternative fuel infrastructure hampers the market introduction of vehicles using alternative fuels and delays their environmental benefits,"(EU Directive [1]). As of 2018, estimated number of natural gas vehicle is represented in Table 4.1.

Biomethane is also a powerful weapon against climate change. In Europe, the demand for transport resulted in a 19.4% surge of GHG emissions in the whole transport sector between 1990 and 2013. As mentioned in previous chapters, biogas system captures methane from anaerobic digestion of organic matters in wastes and crops. Methane emission from natural decomposition of these organic wastes contains up to 25 times Global Warming Potential (GWP) compared with CO_2, (Intergovernmental Panel on Climate Change [2]). Therefore, the CO_2 emissions from burning biomethane are a small fraction of the total global warming potential from the directly emitted methane. As a result, the total carbon footprint is very low, when compared with its fossil fuel equivalents. In fact, its carbon footprint can even be negative ([3]) meaning that driving your biomethane vehicle actually benefits the planet. In addition to using pure biomethane in vehicles, a smart cost-efficient way to reduce GHG emissions is by blending it with natural gas. Blending the two gases,

S. Koonaphapdeelert et al., *Biomethane*, Green Energy and Technology,
https://doi.org/10.1007/978-981-13-8307-6_4

Table 4.1 Estimated number of CNG vehicles in Asia Pacific and top ten CNG countries worldwide (July 2018)

Location	Number of vehicles	Number of filling stations
Globally	~22.6 million	~26,900
Asia Pacific	~12.4 million	~13,600
Total EU	~1.3 million	~3,400

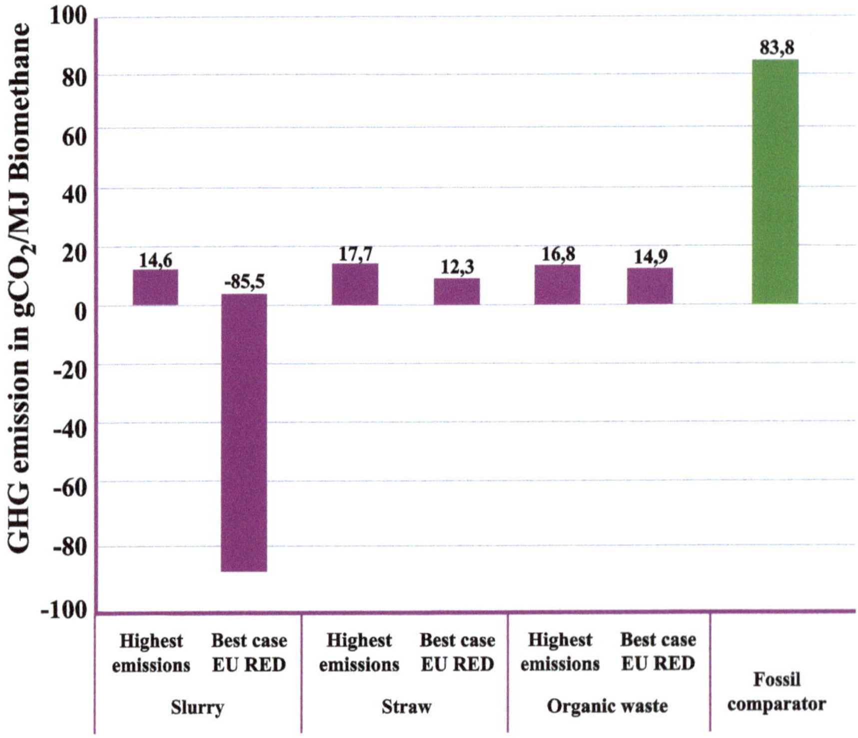

Fig. 4.1 Summary of the GHG emission calculations including scenarios with the highest and lowest emissions for each waste type (Adapted with permission from [3])

even by using a low biomethane-to-natural gas ratio, can result in fuel that has substantially lower emissions than plain natural gas. For example, using a blend with 20% biomethane can yield GHG emission savings of 39% when compared to gasoline on the well-to-wheel basis [4]. This is particularly the case when biomethane from waste with very low (or even negative) GHG emissions is used, see Figs. 4.1 and 4.2. Policies for encouraging the use of biomethane can be integrated with those for electric vehicles to promote an integrated, holistic approach to future transportation. To summarize the advantages of using NG and biomethane in transportation applications.

1. Natural gas causes less pollution compared to gasoline, diesel, and LPG.
2. NG is cost competitive compared to gasoline, diesel, and LPG.
3. NG is a safer gas than LPG. It is lighter than air so if it leaks it rises out of harms way. It is more difficult to ignite compared to other fuels.

Fully upgraded biomethane is comprised solely of methane molecules. Therefore, it has the same properties as natural gas and can be blended, stored, and transported in the same way. Similar to natural gas, biomethane requires a distribution infrastructure, see Table 4.2. There are two principal ways of supplying vehicles with biomethane:

1. Dedicated retail filling stations supplying only biomethane (compressed or liquefied), which often comes from a plant nearby or is shipped in liquefied form (similar to LNG). A dedicated compressed biomethane filling station produced from starch factory waste located in Korat, Northeastern Thailand is illustrated in Fig. 4.3.
2. Fueling stations are connected to the gas grid and offer na
tural gas blended with biomethane. It is impossible to distinguish biomethane from natural gas once it is injected into the grid, but by means of independent and reliable documentation systems the biomethane volumes can be virtually traced and mass-balanced. This allows consumers to buy the equivalent of biomethane being injected by producers. An alternative is to create a legal framework by setting a mandatory biomethane percentage (or blend) for fueling stations as it

Table 4.2 Biomethane production and infrastructure in Europe [5]

Country	Number of biomethane filling stations	Type of biomethane filling station (pure biomethane or blended)	Number of CNG filling stations	Tariff (FiT) or premium (P) for Biomethane (Euro/MWh)
Austria	3	Pure	180	8 (FiT)
Denmark	n/a		7	18.8 (P)
Finland	24	Pure	25	n/a
France	n/a		310	129.7 (FiT)
Germany	308	165 Pure	920	n/a
Hungary	1	Pure	19	n/a
Italy	n/a		1,040	150 (FiT)
Luxembourg	n/a		7	n/a
The Netherlands	60	Pure	141	1.03 Euro/Nm^3
Spain	n/a		69	n/a
Sweden	218	Blended	218	n/a
Switzerland	137	Blended	137	n/a
UK	n/a		8	70 (FiT)
Thailand	4	3 Pure	510	n/a

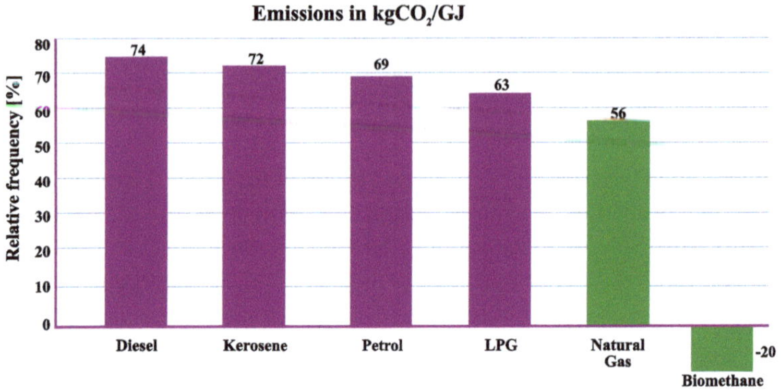

Fig. 4.2 Transportation fuel CO2 Emissions per unit energy produced (Adapted from IPCC Guidelines for National Greenhouse Gas Inventories [2])

Fig. 4.3 RBF; 3 ton per day, dedicated compressed biomethane (CBG) filling station produced from waste in Korat, Thailand

is currently done for liquid biofuels in several countries. This would require the active support of national decision-makers.

4.2 Natural Gas Quality for Transportation

Before substituting biomethane for natural gas in CNG vehicles, it is important to ensure that the gases are compatible. To this end, national governments and standards organizations develop regulations to ensure safety of end use. These regulations are based on the quality of natural gas commonly used in that country. There is not a uniform global standard as there is not a uniform global natural gas quality. Engine manufactures then proceed to manufacture engines to meet the local gas standards. If end applications for natural gas have well-defined properties, it is possible to substitute biomethane for those applications where it meets these requirements. For example, in Thailand, the Department of Energy defined requirements for the use of

Table 4.3 Specifications on the nature and quality of biomethane for vehicles. Announced by Thailand's Department of Energy in 2018

Specification	Value	Test method
1. Dew point 20,000 kPa	4.4 °C (max)	ASTM D 1142
2. Hydrocarbon dew point at a pressure of 4500 kPa which condenses into not more than 1% liquid	10.0 °C (max)	ASTM D 1945 Calculated from equation of State and GRI method
3. Methane number	65 (At least)	
4. Hydrogen (% by Vol.)	0.1 (max)	
5. Carbon dioxide (% by Vol.)	15 (max)	
6. Oxygen (% by Vol.)	1 (max)	
7. Wobbe index (MJ/m^3)	Between 39 and 44	ASTM D 3544
8. Hydrogen sulfide (mg/m^3) 9. Sulfur (mg/m^3)	23 (max) 50 (max)	ASTM D 5504
10. Total volatile silicon compounds (mgSi/m^3)	0.3 (max)	EN ISO 16017-1
11. Amines (mg/m^3)	10 (max)	VDI 2467 Blatt 2

biomethane in vehicles, see Table 4.3. This standard is based on the natural gas that is available in the Thai market. If biomethane has these properties, it is possible to use it as a CNG replacement. The European equivalent standard is shown in Table 4.4 [6]. In Table 4.3, gas properties 1–9 are identical to the natural gas requirements while the total volatile silicon compounds and amines are specific to biomethane. These particular standards will vary from country to country.

4.3 Natural Gas Vehicles

Biomethane that meets a country's natural gas standard, see Tables 4.3, 4.4, can be used in Natural Gas Vehicles (CNGs). The performance will be identical and a driver would not be able to tell the difference. This section will give some background on CNGs and some of the issues and policies needed for adaption on a large scale (Fig. 4.4).

CNG vehicles, such as cars, trucks, vans, and buses, can be divided into two types ([7]):

1. Dedicated CNG: Vehicles using NG only have engines designed and developed for NG use only. These run on an OTTO thermodynamic cycle similar to an IC engine. They utilize spark plugs for ignition and operate at slightly higher pressures than gasoline engines. These vehicles emit less particulates and soot. The cost of these engines is about 20–30% over regular gasoline engines, depending on the type of car and the manufacturer. They can also be converted from gasoline engines

Table 4.4 Requirements, limit values, and related test methods for natural gas and biomethane as automotive fuels

Parameter	Unit	Limit values		Test method (Informative)
		Min	Max	
Total volatile silicon (as Si)	mgSi/m^3		0.3	EN ISO 16017-1:2000 TDS-GC-MS
Compressor oil		The biomethane shall be free from impurities other than "de minimis" levels of compressor oil and dust impurities		ISO 8573-2:2007
Dust impurities		The biomethane shall be free from impurities other than "de minimis" levels of compressor oil and dust impurities		ISO 8573-4:2001
Hydrogen	(% mol)		2	EN ISO 6974-3 EN ISO 6974-6 EN ISO 6975
Hydrocarbon dew point temperature (from 0.1 to 7 MPa absolute pressure)	(°C)		–2 (as in EN 16726)	ISO 23874 ISO/TR 11150 ISO/TR 12148
O_2	% mol		01	EN ISO 6974-series EN ISO 6975
Hydrogen sulfide + Carbonyl sulfide (as sulfur)	(mg/m^3)		5 (as in EN 16726)	EN ISO 6326-1 EN ISO 6326-3 EN ISO 19739
S total (including odorization)	(mgS/m^3)		30	EN ISO 6326-5 EN ISO 19739
Methane number	Index	65 (as in EN 16726)		Annex A of EN 16726:2015
Amine	mg/m^3		10	VDI 2467 Blatt 2:1991–08
Water dew point	Class A	–10 °C at 20000 kPa		ISO 6327 (applicability at 20 000 kPa)
	Class B	–20 °C at 20000 kPa		
	Class C	–30 °C at 20000 kPa		

using an aftermarket installation kit. They are typically used in smaller vehicles such as cars and light vans.

2. Diesel Dual Fuel (DDF): Vehicles using this system operate with two fuels, NG and diesel. They operate on the diesel thermodynamic cycle with pressure ignition. The natural gas enters the engine with the intake air. The diesel is injected and causes combustion. Less diesel is needed, up to 60% and less emissions enter the environment. This results in fuel cost savings between 15 and 40%. This method does not have to modify the original diesel engine, but only install CNG equipment to the air intake system. The NG is mixed with air prior to entering the engine (Fig. 4.5).

There are two gas mixing systems:

(a) CNG Cars and Vans Filling up at CNG Station (Editorial credit: Deshzx / Shutterstock.com)

(b) CNG Truck with Multiple High Pressure Tanks for Long Distance Hauling

(c) CNG Bus as Low Emission Public Transport for the City (Editorial credit: meowKa / Shutterstock.com)

(d) CNG filling station (Courtesey of ERDI-CMU)

Fig. 4.4 Selected photos of CNG vehicles and filling stations

Fumigation: Operates similarly to a carburetor. A pressure regulator controls the flow of natural gas to the air. The flow rate of the gas changes proportionally with the air intake flow rate. The control system is a mechanical open-loop system. Some systems have an Electronic Control Unit (ECU) to control the NG supply into the combustion chamber of the engine in proportion with the speed of the engine.

Injection: This system directly injects the gas into the air intake manifold. It is a closed-loop system using an exhaust oxygen sensor to provide feedback on the optimal mixture ratios (Fig. 4.5).

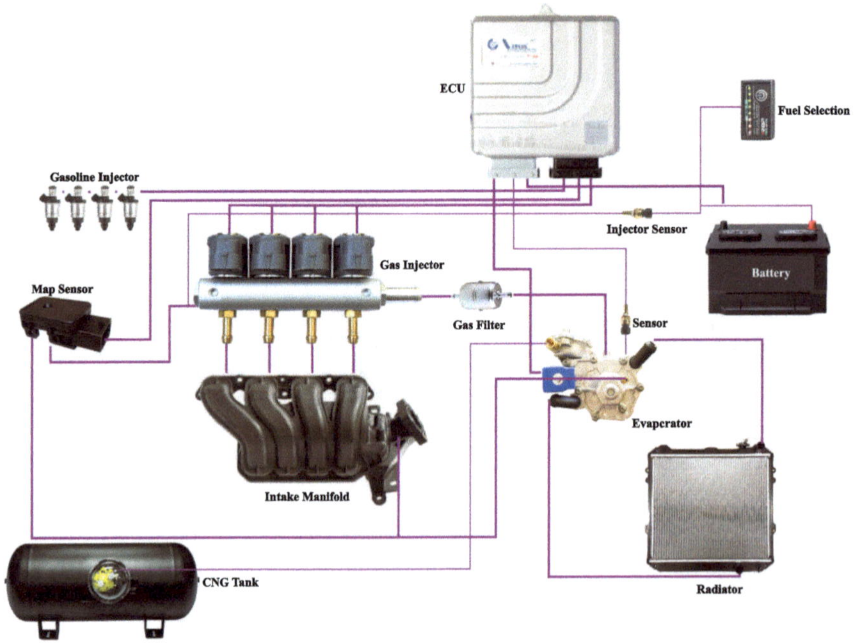

Fig. 4.5 Diagram of natural gas or biomethane injection system for vehicles

4.3.1 CNG Development Obstacles

There are reasons why large-scale adaption of CNGs has not yet happened. It does suffer from some disadvantages as follows:

1. The compressed NG tank is bulky and heavy due to the high pressures required (up to 25 MPa). This affects trunk space, vehicle weight, and balance. The equipment used to install NG is more sophisticated than normal fuels and require greater safety prevention measures.
2. The fuel filling time is longer than gasoline and diesel but still significantly quicker than electric charging. The traveling range of NG per fill-up is significantly shorter than other fuels.
3. There are far fewer compressed NG filling stations as they are limited to locations close to a pipeline network. Off-grid stations are technically possible but require high investment costs.

These issues limit the growth of CNGs. In remote areas with no pipeline, a dedicated biomethane plant may be a solution. It could provide a stand-alone natural gas fuel refilling service. It is cost competitive compared with transporting CNG or Liquefied Natural Gas (LNG) over large distances.

(a) Front view (b) Rear view

Fig. 4.6 Small-scale biomethane water scrubbing system

4.4 Case Study: CBG versus CNG

For the purposes of testing a local biomethane filling station, a small-scale water scrubbing system was developed to supply the fuel, see Fig. 4.6. This system can provide 10 kg/h and is located in a standard size (40 ft) container. This allows it to be shipped easily on a truck bed. The raw biogas is obtained from the waste of a local swine farm. This system was supported by the Thai Ministry of Energy and implemented in Chiang Mai University in 2013.

A new Mitsubishi Triton-CNG pickup truck was used to compare the performance of biomethane with CNG and gasoline (gasohol 91). It is shown in Fig. 4.7. The biomethane was obtained from the station shown in Fig. 4.6. The testing comprised of three components.

1. The truck was mounted on a dynamometer and the torque and horsepower were measured for all three fuels.
2. The truck was driven around the locale and the fuel economy was measured.
3. The emissions, carbon monoxide and hydrocarbons, were measured for all fuels.

Three different percentages of methane in the biomethane were tested, 83, 85, and 90% by volume, respectively. The biomethane always met the Thai standard for vehicle as shown in Table 4.3.

Fig. 4.7 CNG vehicle used in fuel comparison testing

4.4.1 Results

The horsepower and torque results are shown in Fig. 4.8. The results show that gasoline provides about 10% more power across all engine operating speeds. This is unsurprising and common to all CNGs. The gasoline had a maximum horsepower of 104 hp (77.5 kW) at 4,500 rpm and a maximum torque of 175 Nm at 2,500 rpm. The maximum horsepower of the biomethane was 87 hp (64.9 kW) at 4,500 rpm with a maximum torque of 155 Nm at 2,500 rpm. The important result is the performance of biomethane compared with natural gas. They are identical. Even at lower methane percentages, the results were still the same. If the engine firing timing was adjusted then performance differences in the biomethane grades were observed. However, at the factory default setting, there was no observable difference.

Fuel consumption tests were conducted by driving the truck around the roads of Chiang Mai at an average speed of 70–80 km/h. The cost of gasoline at the time of testing was $1.15 per liter and the NG was $0.37/kg. The gasoline obtained a fuel consumption of 11.33 km per liter or $0.1/km. The CNG fuel consumption was 13.76 km/kg or $0.03/km. The 85% biomethane fuel consumption was 13.76 km/kg or $0.027/km. These conditions are applicable when the air-conditioning system is turned off, and when it was turned on, the fuel efficiency decreased slightly as shown in Table 4.5.

The level of carbon monoxide (CO) was measured for all three fuels. Gasoline produced the highest amount of carbon monoxide, from 0.1 to 0.13%. The CO emissions from biomethane and CNG were approximately three times less across all engine speeds ranging from 0.03 to 0.05% of the exhaust volume.

The hydrocarbon (HC) emissions from CNG and gasoline were between 115 and 135 ppm. Those from biomethane were 95–110 ppm, slightly less than CNG. This is because, in biomethane, there are no longer chained hydrocarbons, ethane, and propane that are found in natural gas.

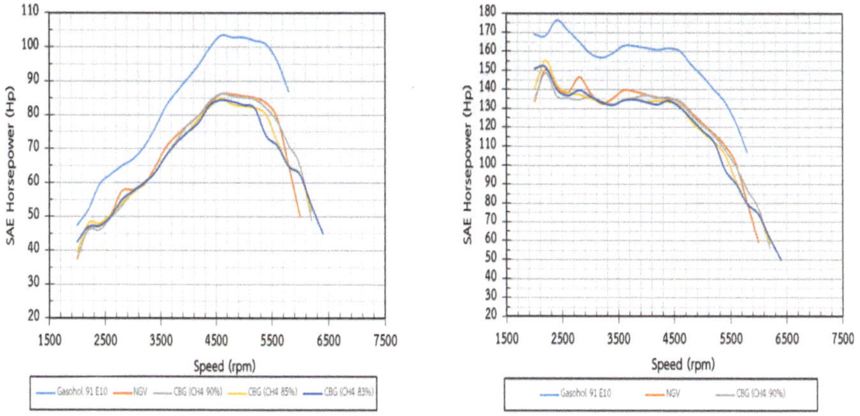

(a) Horsepower (b) Torque

Fig. 4.8 Horsepower and torque for three different fuels

Table 4.5 Fuel economic performance for gasoline, CNG, and biomethane

Driving conditions	Fuel	(km)	Fuel consumed	Specific fuel consumption
Air con. turned off	Gasohol 91	51.0	4.50 L	11.30 km/L ($0.10/km)
	CNG	51.0	3.71 kg	13.76 km/kg ($0.03/km)
	Biomethane (85%)	51.0	3.68 kg	13.79 km/kg ($0.03/km)
Air con. turned on	Gasohol 91	51.0	4.75 L	10.74 km/L ($0.10/km)
	CNG	51.0	3.99 kg	12.77 km/kg ($0.03/km)
	Biomethane (85%)	50.6	3.86 kg	13.14 km/kg ($0.03/km)

All other biomethane emissions including SOx, NOx, and particulates were all well under emission regulations. They were measured but not described in detail here.

References

1. European Parliament and Council (2014) On the deployment of alternative fuels infrastructure, October 2014. EU Directive 2014/94/EU
2. Penman J, Gytarsky M, Hiraishi T, Irving W, Krug T (2006) IPCC guidlines for national greenhouse gas inventories. Technical report, Intergovernmental Panel on Climate Change
3. Majer S, Oehmichen K, Kirchmeyr F, Scheidl S (2016) Calculation of ghg emission caused by biomethane. Technical report, BIOSURF
4. Europe NGVA (2015) Report of activities 2014–2015. Technical report, Natural and Bio Gas Vehicle Association
5. EBA (2018) Biomethane in transportation. Technical report, European Biogas Association
6. European Committee for Standardization (2017) Natural gas and biomethane for use in transport and biomethane for injection in the natural gas network - part 2: automotive fuels specification

7. PTT Public Company Limited (2019) NGV installation. http://www.pttplc.com/en/Products-Services/Consumer/For-Vehicle/NGV/PTT-NGV/Pages/Installation.aspx
8. EEA Report (2014) Air quality in Europe: 2014 report. Technical Report No 5/2014, European Environment Agency

Chapter 5
Biomethane in Domestic and Industrial Applications

5.1 Liquid Petroleum Gas

In this chapter, Thailand will be used as a case study for replacing LPG with biomethane. In 2018, Thailand used around 19,100 tons of LPG every day. Approximately, 31% of this was for domestic cooking and 10% was for industrial use. The remaining uses were in the transportation and agricultural sectors. Figure 5.1 displays the annual LPG demand and the annual LPG used in cooking and industrial applications. In 2018, approximately 2.1 million tons were used for domestic cooking purposes. The Thai government used to subsidize the LPG price but has been planning to remove these subsidies. According to the EFIA, an independent public organization under the Energy ministry, the LPG subsidy at the start of 2014 was costing the Thai government $3 million (฿100 million) per day. Having an artificially low price did not promote energy conservation and encouraged cross-border smuggling as Cambodia and Laos have higher LPG prices. Having a viable alternative to LPG, such as biomethane, would allow for higher LPG prices without as many adverse consequences. There are additional benefits to using the renewable biomethane from increasing the incomes of farmers who produce the biomethane from farm waste, reducing the cost of importing petroleum products and reducing greenhouse gas emissions.

The economic potential of biogas is discussed in depth by [2, 3]. Chaiprasert [4] and Aggarangsi et al. [5] give a summary on the state of biogas production from agricultural waste in Thailand. Nasir et al. [6] and Sakar et al. [7] provide a review on the present state of the existing technology for producing biogas from agricultural waste. Chapter 3 of this book and [8] review the production of biomethane from biogas.

S. Koonaphapdeelert et al., *Biomethane*, Green Energy and Technology,
https://doi.org/10.1007/978-981-13-8307-6_5

Fig. 5.1 LPG total demand and use in domestic and industrial applications for Thailand, from [1]

5.2 Biomethane for Domestic Cooking

Bond and Templeton [9] give a historical picture of biogas use and review the use of biogas in domestic cooking in South Asia. For domestic cooking purposes, a biomethane methane content of 85% is deemed minimal. Other applications such as transportation and gas grid injection require a higher methane content. However, domestic cooking is not as critical an application as automotive power or pipeline gas quality. Regardless of the methane content, it is still important to remove the hydrogen sulfide as this can corrode storage tanks and connecting pipes, as discussed in Chap. 2. Dai et al. [10] studied biomethane flames in burners, with a maximum methane percentage of 70% and a controllable inlet air supply.

For domestic cooking applications, the biomethane can be stored in a glass fiber tanks, or any suitable gas storage tank, see Fig. 5.2. These tanks are designed for a pressure of 20 MPa and have a 15 year life span. They are made from glass fiber and epoxy resin. The tank output is regulated down to the required stove pressure.

5.2.1 Domestic Stoves

In South and Southeast Asia, there are two basic styles of stove for domestic cooking purposes. They are shown in Fig. 5.3. They are named after the way the fuel/air mixture flows through the stove. The path the air and fuel takes through both swirl and radial burners is identical apart from the exit head geometry. In a swirl burner (SB), which is also the most common type, the fuel mixture swirls outward at an angle, see Fig. 5.4. The radial burner (RB) is a more direct style with the combustion mixture exiting in the vertical direction. They both have a small inner flame when

Fig. 5.2 Biogas storage tank, regulator, and stove (Reprinted with permission from [11])

the stove is set to the low setting and a larger outer ring when the setting is medium or high and both are designed for use with LPG. Previous research on the efficiency of each burner showed that the swirl burner had a maximum efficiency of 65% on a fuel LHV basis, which was 5–15% more efficient than the radial burner [12, 13]. Jugjai et al. [14] ran efficiency tests on 380 LPG stoves and found an average efficiency of 49% across a range from 35 to 65%. Lucky and Hossain [15], in a similar fashion, examined domestic natural gas stoves in Bangladesh and found approximate efficiencies which varied from 40 to 59%.

Both swirl and radial burners operate similarly. An LPG tank supplies the fuel, which passes through a regulator with an output pressure of $5\,\text{kPa}_{gauge}$. The fuel then flows past the adjustable nozzle and through either 1 or 2 nozzles depending on whether the setting is low or high. Air is entrained at this point and the fuel/air mixture enters the burner head, see Figs. 5.4 and 5.5. There is a smaller inner burner head for low heating rates and a larger outer burner head for medium and high heating rates.

At the low valve setting, the fuel only flows through nozzle 1, in Fig. 5.4. At medium and high valve settings, the fuel flows through both nozzles. At the nozzle exit, outside air is entrained in a similar way to a jet pump operation, see Fig. 5.5.

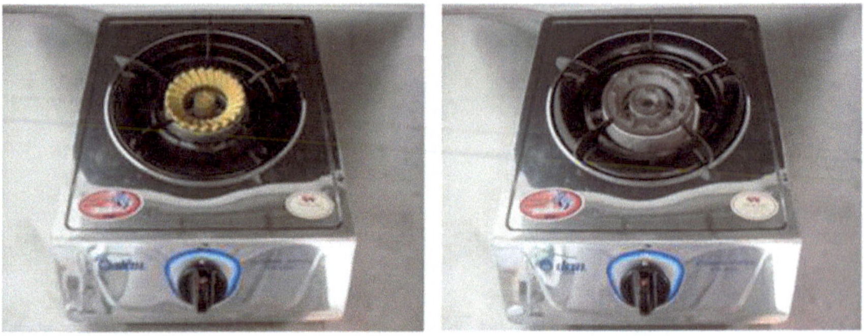

Fig. 5.3 Swirl and radial cooking stoves (Reprinted with permission from [11])

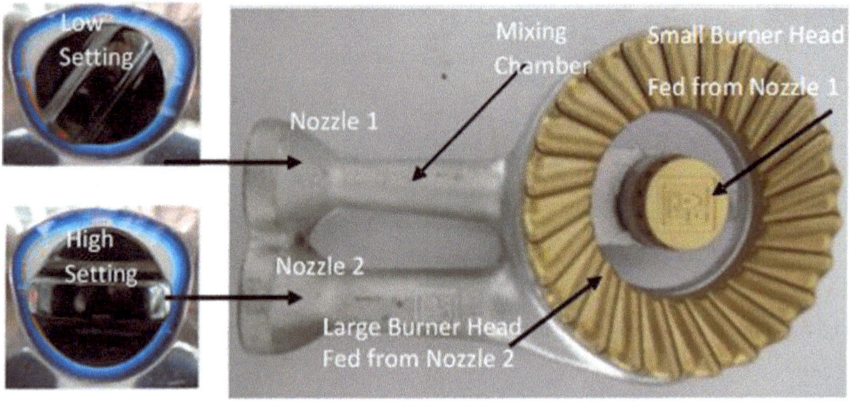

Fig. 5.4 Swirl burner head and inlet manifold (Reprinted with permission from [11])

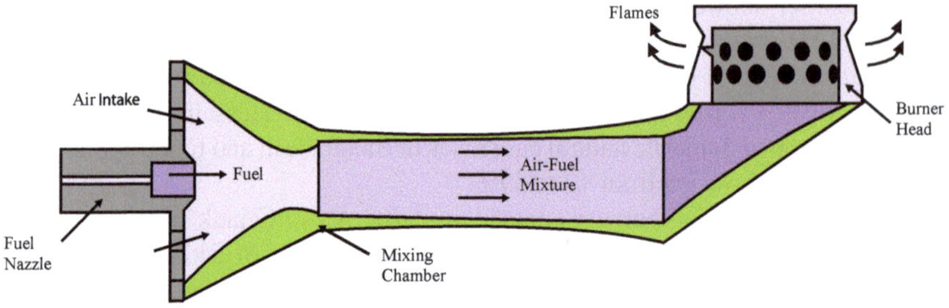

Fig. 5.5 Cutaway view of fuel/air flow through the mixing chamber (Reprinted with permission from [11])

Some differences between biomethane and LPG are their energy content and thermodynamic state when stored. The biomethane is a mixture of 85% methane and 15% carbon dioxide with a lower heating value (LHV) of 33.7 MJ/kg. This is lower than that of both LPG (46 MJ/kg) and pure methane (50 MJ/kg). Since LPG

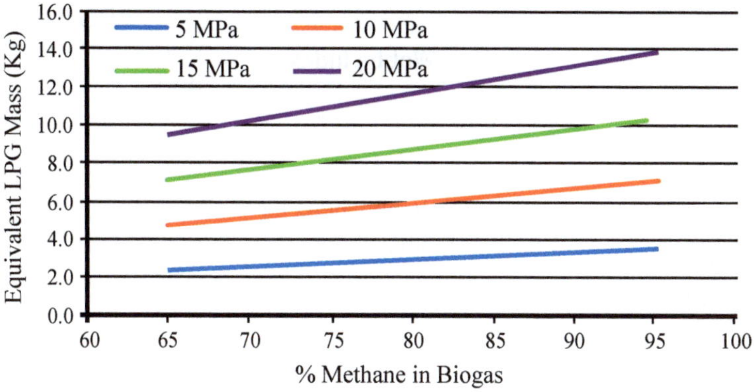

Fig. 5.6 Energy content in a 100 L tank of biomethane

liquefies at a relatively low pressure (700 kPa) while methane cannot be liquefied at room temperatures, the volume of a tank of methane will be larger than an LPG tank, for the same energy stored. The energy content of a 100 L tank filled with biomethane will increase with pressure and the percentage of methane in the gas. A 1 m long, 34 cm diameter, 100 L tank is normally used for compressed natural gas in light vehicles. Figure 5.6 shows the energy content for this tank for different methane contents and tank pressures. At 10 MPa and 80% methane, the energy content is 267 MJ which is equivalent to a 5.8 kg tank of LPG (1 kg LPG ~ 2 L). At 15 MPa and 92% methane, the energy content is 462 MJ which is equivalent to a 10 kg tank of LPG. Even at 20 MPa and 100% methane, the energy equivalent is only 14.5 kg of LPG which is a little less than the standard 15 kg LPG bottle size. Going above 20 MPa is not advisable for safety considerations. It is also the maximum pressure limit for natural gas vehicles in Thailand.

5.3 Fuel Switching Analysis

An analytical design procedure for modifying domestic stoves to safely combust biomethane is presented here. It becomes necessary to change the stove nozzle diameters and fuel supply pressure to get a stable flame when switching from LPG to biomethane. The mixing chambers and burner heads remain the same, see Fig. 5.4. First, an analysis of the Wobbe index for both fuels is undertaken.

5.3.1 Wobbe Index

The Wobbe Index (WI) or Wobbe number is an indicator of fuel gas interchangeability. It is used to compare the combustion energy output of different composition fuel gases. The higher a gas's' Wobbe number, the greater the heating value for a

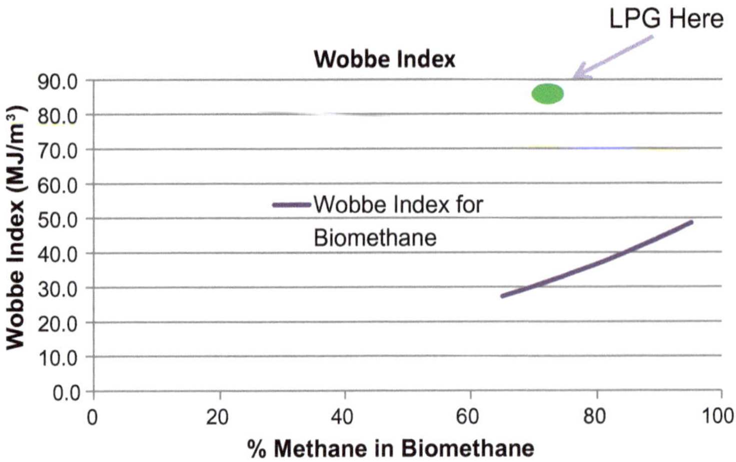

Fig. 5.7 Wobbe Index for biomethane with varying methane content (Reprinted with permission from [11])

quantity of gas that flows through an orifice in a given time period. The Wobbe Index for LPG is around 85 MJ/m^3 and for biomethane its value depends on the percentage of methane as shown in Fig. 5.7. For 85/15 biomethane, its Wobbe index is 40 MJ/m^3. This is less than half that of LPG meaning that these fuels are not directly interchangeable.

The Wobbe Index, WI, is calculated from the ratio of the volumetric heat, H_v (J/m^3) per cubic meter divided by the square root of the specific gravity, s, which is the ratio of the gas density (kg/m^3) to air density (kg/m^3) as shown in Eq. 5.1.

$$WI = \frac{H_v}{\sqrt{\rho_{gas}/\rho_{air}}} = \frac{H_v}{\sqrt{s}} \tag{5.1}$$

The energy flow through a nozzle will be proportional to the product of the rate of gas flow $\dot{V}$ (m^3/s) and volumetric heat H_v (J/m^3), see Eq. 5.2.

$$\dot{E} = \dot{V} H_v \tag{5.2}$$

The mass flow rate of the gas can be calculated from the continuity equation, where the gas velocity can be determined, v_{gas} (m/s). The gas velocity is thus related to the orifice pressure drop through the Bernoulli equation. Neglecting the effect of gravity on the flow, Eq. 5.3 is derived for the gas flow.

$$\dot{V} = A v_{gas} = A\sqrt{\frac{2\Delta P}{\rho_{gas}}} \tag{5.3}$$

On combining Eqs. 5.3 and 5.2, it gives

$$\dot{E} = A\sqrt{\frac{2\Delta P}{\rho_{gas}}} H_v \quad (5.4)$$

The heat released per unit time depends on the volumetric heating value of the gas, the gas density, orifice area, and orifice pressure drop. Substituting Eq. 5.1 for the Wobbe Index gives

$$\dot{E} = A\sqrt{\frac{2\Delta P}{\rho_{air}}} WI \quad (5.5)$$

From Eq. 5.5, the Wobbe Index is proportional to the energy flow through a given nozzle of a certain size. If two different gases have the same Wobbe Index they will produce the same heating rate though any given nozzle. The WI is one of the key variables in comparing gas interchangeability.

5.3.2 *Flame Stability*

An important parameter controlling the combustion behavior of gaseous fuel is the laminar burning velocity (S_L). A flame is reported to be stable when there is a balance between the reactant's velocity and the laminar burning velocity. When an imbalance between both velocities occurs, instability phenomena such as blowoff and flashback appear. In a domestic burner, if the laminar burning velocity is lower than the speed of the incoming unburnt reactants, the flame may detach from the burner (known as blowoff). On the other hand, if the laminar burning velocity is higher than the flow rate of the unburnt reactants, there is a possibility that the flame may fall back into the premixing section of the burner in a phenomenon known as flashback.

The laminar burning velocity values of biogas and others are shown in Table 5.1. The laminar burning velocities of LPG and biogas are small relative to those of conventional fossil fuels. This is not particularly good news since it means that high mixture flow rates will result in blowoff. A low laminar burning velocity means that small flow rates are necessary to achieve stability.

Therefore, switching fuels from LPG to biomethane may induce a flame stability problem because of different laminar burning velocities. Prior to using biomethane in domestic stoves, it is necessary to carry out experiments to ensure flame stability. The end goal is to ensure a smooth, safe, and reliable combustion operation.

Table 5.1 Comparison of laminar burning velocity for different fuels (Reprinted with permission from [11])

	Fuel Type			
	LPG [16]	Biogas [17]	Natural gas [18]	Producer gas [19]
ER range	0.8–2.0	0.6–1.2	0.8–1.5	0.6–1.7
SL range (cm/s)	13–40	7–26	10–37	7–29
max SL at ER (cm/s)	40@1.3	26@1.0	37@1.1	29@1.2

Table 5.2 Thermodynamic properties of both the biomethane and LPG

Fuel	Biomethane	LPG
Constituent 1	85% Methane	60% Propane
Constituent 2	15% Carbon dioxide	40% Butane
Pressure (kPa)	106.325	106.325
Temperature (K)	298	298
Gas constant (kJ/kg K)	0.4	0.17
Density (kg/m^3)	0.9	2.1
Lower heat value (MJ/kg)	33.7	46.1
Higher heat value (MJ/kg)	37.4	50.3

5.3.3 *Nozzle Diameters*

In order to substitute biomethane for LPG, changes are made to the fuel/air intake system. Specifically, the nozzle diameters are modified along with the fuel supply pressure. To estimate the required nozzle diameter, a thermodynamic analysis is carried out. The thermodynamic properties used in this analysis are shown in Table 5.2.

The LHV of pure methane is 50 MJ/kg. Since we are dealing with 85/15 biomethane there are 5.6 times more moles of methane than carbon dioxide, $\frac{\dot{n}_{CH_4}}{\dot{n}_{CO_2}} = 5.66$, where $\dot{n}$ is the molar flow rate. Given

$$\dot{m}_{biomethane} = M_{CH_4}\dot{n}_{CH_4} + M_{CO_2}\dot{n}_{CO_2} \tag{5.6}$$

$$\frac{\dot{m}_{biomethane}}{\dot{m}_{CH_4}} = 1 + \frac{M_{CO_2}\dot{n}_{CO_2}}{M_{CH_4}\dot{n}_{CH_4}} \approx 1 + \frac{(44)(15)}{(16)(85)} = 1.48 \tag{5.7}$$

So for every 1.48 kg of biomethane that enters the stove, 1 kg is pure methane. This gives a calculated lower heating value for biomethane of $LHV_{biomethane} = \frac{LHV_{CH_4}}{1.48} = 33.7\,\mathrm{MJ/kg}$. For equal heating rates, the lower density and heating value for the biomethane require a higher mass flow rate through the nozzle. The higher exit velocity entrains too much air into the intake manifold and results in flame blowoff.

The solution is to increase the diameter of both stove nozzles to allow more mass flow through while keeping the fuel velocity at the nozzle exits relatively constant.

The assumption in the following analysis is that for a given heating rate the mass flow of fuel is known, see Eq. 5.10. The mass flow of biomethane is higher than LPG for an equivalent heating rate. Since the intake manifold and nozzle assembly are optimized for LPG, the goal is to find nozzle diameters that give the same average exit velocity for biomethane. This is not an exact requirement but rather a guideline for estimating the nozzle diameter. Experiments can then be performed around this estimate. Let nozzle 1 have a diameter of D_1 and an exit average velocity of V_1. D_2 and V_2 then apply to the second nozzle, nozzle 2. The original nozzle 1 diameter is $D_1 = 0.6$ mm and nozzle 2 diameter is $D_2 = 0.9$ mm, which are standard diameters used with LPG. Changing the diameter of either nozzle affects the fuel exit velocity. The pressure drop through both nozzles should be equal, resulting in equal mass flow rates through both nozzles:

$$\frac{D_2^2}{D_1^2} v_1 = v_2 \tag{5.8}$$

At the lowest stove setting, the output heating rate was around 0.5 kW. At the highest setting, the heating rate was 4 kW. Taking 0.5–4 kW as the heating range, the nozzle exit velocities are plotted. These are shown in Fig. 5.8 for both nozzles. They are plotted in pairs as changing one nozzle diameter affects the other.

The dashed line in Fig. 5.8 is the LPG baseline. At 2 kW, the LPG velocity through nozzles 1 and 2 is 12 m/s and 27.5 m/s, respectively. Using the same sized diameters for biomethane, the corresponding velocities are 38 and 86 m/s. These are triple the LPG velocities and would lead to unstable or nonexistent flames from blowoff.

Enlarging both nozzle diameters reduces the average velocity. The set that results in average velocities, most similar to the LPG, is nozzle 1 having a $D_1 = 1$ mm diameter and nozzle 2 having a 1.8 mm diameter. This gives a theoretical point to begin modifying the stove.

5.3.4 Pressure Drop

For LPG, the standard supply pressure is 5 kPa_{gauge}. The equivalent pressure drop for biomethane is modeled here. The stove pressure drop occurs in two main sections, the flow adjustment valve and the nozzle. Modeling the flow valve as a restrictor, the mass flow rate through a restriction is given by

$$\dot{m} = Const\sqrt{\rho(\Delta P)} \tag{5.9}$$

where the constant is a function of the restriction geometry. For the same heating rates, the mass flow rate of biomethane is given by

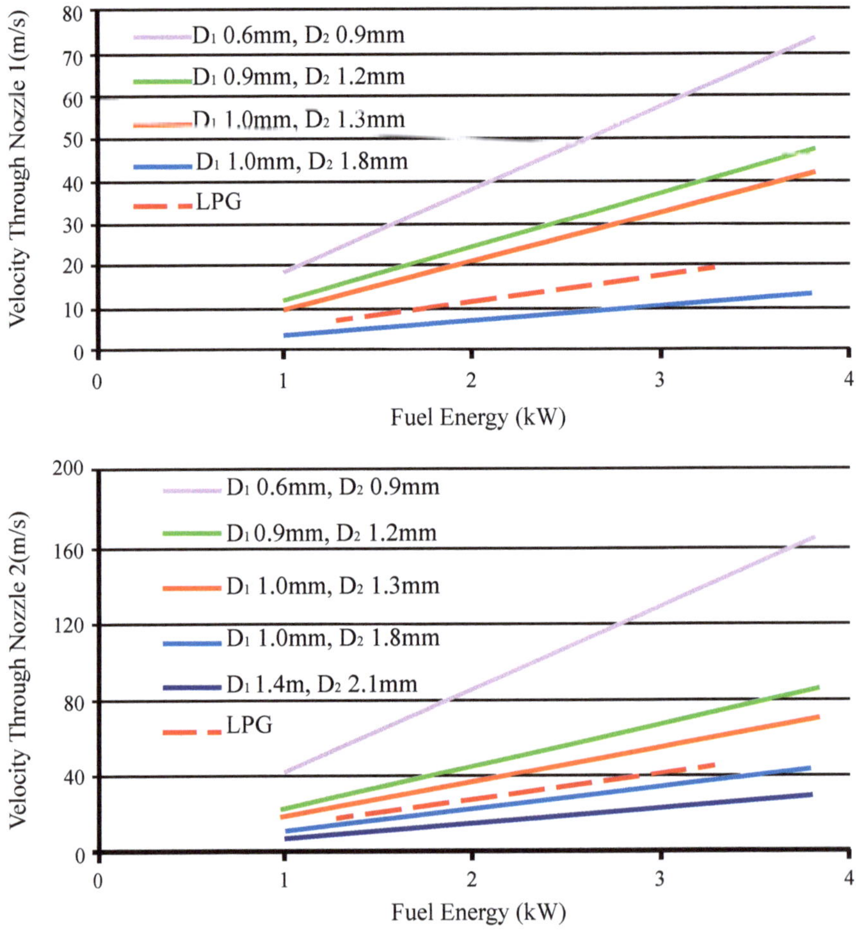

Fig. 5.8 Average exit velocities through nozzle 1 (top) and nozzle 2 (bottom) (Reprinted with permission from [11])

$$\dot{m}_{biomethane} = \frac{\dot{m}_{LPG} LHV_{LPG}}{LHV_{biomethane}} \tag{5.10}$$

Combining Eqs. 5.9 and 5.10 leads to an estimate of the biomethane pressure drop:

$$\frac{\Delta P_{biomethane}}{\Delta P_{LPG}} = \left(\frac{LHV_{LPG}}{LHV_{biomethane}} \right)^2 \frac{\rho_{LPG}}{\rho_{biomethane}} = 4.3 \tag{5.11}$$

Therefore, for equivalent heating rates, the pressure drop for biomethane should be approximately four times that of the LPG. This gives a pressure drop of 20 kPa.

5.3.5 Hardware Changes to the Stoves

The previous analysis suggests that stove nozzles must be enlarged and the supply pressure is increased to substitute biomethane for LPG. Both must happen because if only the supply pressure is increased, then too much combustion air would be entrained. This excess air at the burner head results an unstable or no flame (blowoff). Therefore, the nozzle diameter must also be enlarged. From Fig. 5.8, the nozzle diameter pair of 1.0 and 1.8 mm gave a similar velocity profile to that of LPG.

Suwansri et al. [11], at ERDI, tested biomethane combustibility in domestic stoves at various pressures and nozzle sizes. The result of these experiments is a map of particular pressures and nozzle sets that produce a stable flame at all stove settings and those that did not. These maps are shown for each stove in Fig. 5.9.

Figure 5.9a is a swirl burner and Fig. 5.9b is a radial burner. The x-axis is pressure in kPa. The y-axis is the diameter of the particular nozzle set used. A black circular marker means that it was not possible to get a flame. For example, in Fig. 5.9a, at standard LPG conditions of 5 kPa and a 0.6 mm inner nozzle and 0.9 mm outer nozzle diameter, no biomethane flame was possible. This validates the theoretical arguments presented in Sect. 5.3.3 that it is not possible to directly substitute biomethane for LPG. The LPG condition is shown on each figure just for comparison.

A square marker meant that there was a very low flame. As the fuel flow rate decreases, this flame is very likely to extinguish. The amount of fuel reaching the burner head is insufficient for combustion. This is why the square markers are seen at low pressure and/or low diameters. A diamond marker means that the flame is very high. The term "very high" again is subjective but in this context it means that for the radial stove a 25 mm high flame is too high and 50 mm is too high for the swirl burner. A high flame indicated too much fuel entering the burner head. This is why the diamond markers are observed at high pressures and/or high diameters. A triangular marker is the optimal condition where the flame at all stove settings most closely resembles an LPG flame.

A point was selected in the middle of the triangular markers which is selected as the operating point for the stove to run on biomethane. The point selected is marked with a large square in Fig. 5.9. This condition has a pressure of 20 kPa and a nozzle set of 1.0 and 1.8 mm. Other nozzle sets above and below this should also work. However, this point was selected because it is in the middle of the range and the theoretical analysis predicted this particular size. A pressure of 30 kPa would have also worked, however, a lower pressure is preferred as the risks of gas leaks inside the stove decrease. Figure 5.10 shows pictures of the biomethane flames at the maximum fuel flow rate for the optimal condition described in the previous paragraph. Table 5.3 contains a summary of these optimal conditions for both stoves and compares them with the LPG operating parameters.

The optimal fuel supply pressure is 20 kPa and the optimal inner and outer nozzle diameters were chosen to be 1.0 mm and 1.8 mm, respectively, as previously

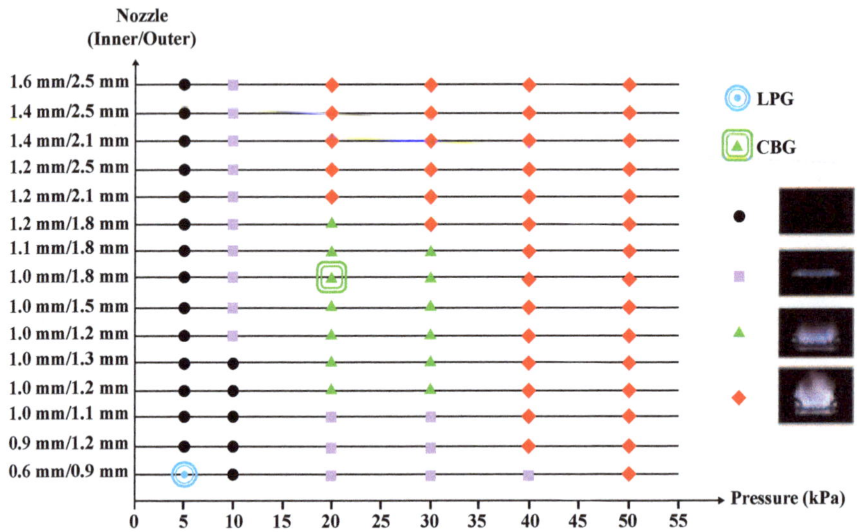

(a) Swirl burner optimal nozzle diameter and supply pressure for biomethane

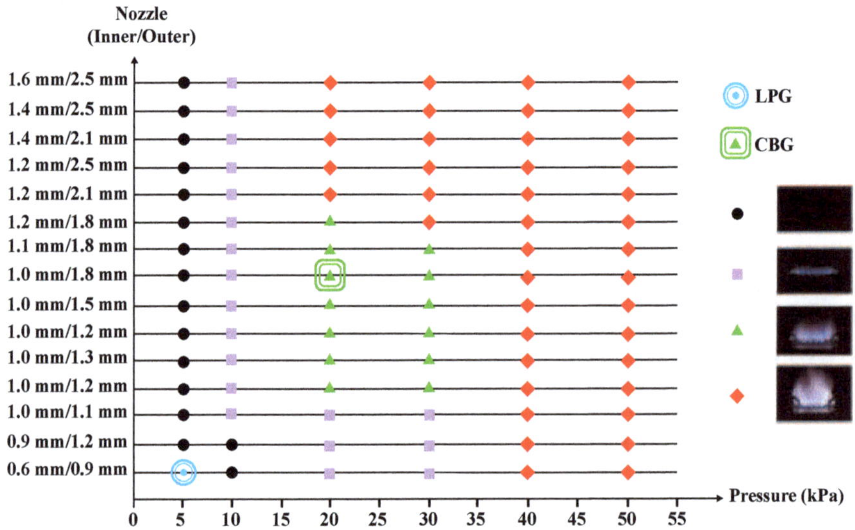

(b) Radial burner optimal nozzle diameter and pressure for biomethane

Fig. 5.9 Radial and swirl burner optimal settings for biomethane, (Reprinted with permission from [11])

predicted. These diameters represent an increase in nozzle area of 178% and 300%, respectively. At these diameters and pressure, the fuel exit nozzle velocity is closest to that of LPG.

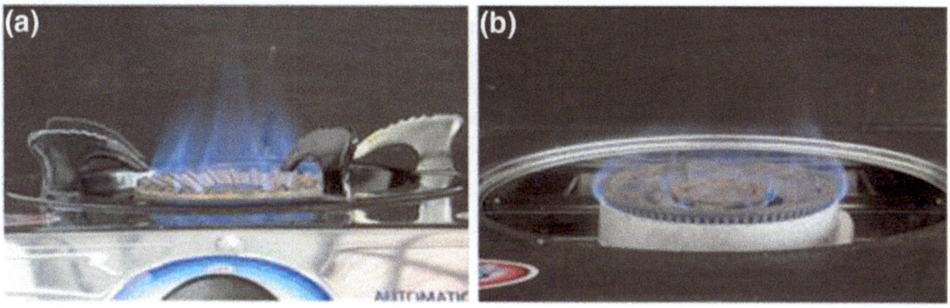

Fig. 5.10 Biomethane combustion at high flow settings for **a** swirl and **b** radial stoves (Reprinted with permission from [11])

Table 5.3 Optimal parameters for biomethane combustion

	Optimal inlet pressure (kPa)	Optimal nozzle inner diameter (mm)	Optimal nozzle outer diameter (mm)
Swirl burner—Biomethane	20	1.0	1.8
Radial burner—Biomethane	20	1.0	1.8
Swirl burner—LPG	5	0.6	0.9
Radial burner—LPG	5	0.6	0.9

5.4 Energy Output and Efficiency

For a domestic stove, the heating time is affected by the energy delivered to the stove head. One method used to determine the input energy is to measure the fuel flow rates with a scale. Suwansri et al. [11] measured stove heating efficiencies according to the standard DIN EN 203-2. This is a European standard covering gas heated catering equipment. It involves heating 7.8 kg of water by 70 °C and measuring the time and quantity of fuel to do so. The efficiency can be obtained from Eq. 5.12.

$$\eta = \frac{m_{water}.C_{water}.\Delta T}{Q_{fuel}.LHV} \tag{5.12}$$

where, in Eq. 5.12,

m_{water} is the initial mass of water in kg,

C_{water} is the specific heat capacity of water in J/kgK,

ΔT is the temperature increase in the water, in °C,

Q_{fuel} is the volume of fuel used in m^3, and

LHV is the lower heating value per unit volume of fuel in J/m^3.

The energy content of the fuel at different settings is shown in Table 5.4 along with the efficiency test results.

Table 5.4 Stove energy output and efficiency at optimal conditions (Reprinted with permission from [11])

	Bio-methane	LPG	Bio-methane	LPG	Bio-methane	LPG	Bio-methane	LPG
	Medium setting (kW)		High setting (kW)		High setting efficiency (%)		High setting heating times (mins)	
Swirl burner	1.03	1.4	2.0	3.2	68	53	29	23
Radial burner	1.4	1.3	2.15	3.2	64	48	26	25

The LPG outputs 26% more energy across all stove settings. This means that when you place the stove dial on a "medium" setting, the energy output will be 26% higher if the fuel used is LPG. A larger nozzle set could be used in order to increase the energy output from the biomethane; however, having less energy does not necessarily mean taking a longer time to heat. The biomethane stoves operate around 15–16% more efficiently than the LPG at the high settings. This indicates that for the high-powered LPG flame output much of the heat is wasted. The LPG efficiency readings correspond well with the report from [14] who tested the efficiency of 380 stoves and found an average efficiency of 49%. For both fuels, the heating times were comparable, the biomethane-powered swirl burner took 6 min longer but there was negligible difference in heating time for the radial burner. Since 7.8 kg is a large quantity of water, it is suspected that in day-to-day cooking the time difference between both fuels would be barely noticeable.

5.5 Substituting Biomethane in Industrial Applications

Similarly to domestic stoves, care must be taken when substituting biomethane for LPG in industrial applications. The flow rates and heating outputs can be an order of magnitude larger. The case study used here will be the Thai ceramics industry. Certain ceramic manufacturing sites in Thailand date back over 5000 years. Ceramic firing can be thought of as a sintering-type process to bond the clay molecules together at high temperatures ~1200 °C. An example of a LPG-powered high-temperature shuttle kiln for ceramic heating is shown in Fig. 5.11.

Liquefied Petroleum Gas (LPG) is extensively used in industrial ceramic kilns. In Thailand, the ceramic industry employs directly and indirectly over 75,000 people generating annual exports in excess of $910 million. In 2018, 687,000 tons of LPG were used in industrial applications. It is estimated that a cost savings of up to 30% can be obtained using biomethane, with a payback period of a little over four and a half years, factoring in the cost of the changeover.

Fig. 5.11 A High-temperature shuttle kiln

5.5.1 *Biomethane Substitution in Shuttle Kilns*

Similarly to a stove, a kiln uses a burner to supply the air and fuel in the correct ratio. The LPG gas is forced through a small nozzle and drags surrounding air with it before the mixture exits the assembly and combusts, Fig. 5.12 shows the most commonly used natural draft burners, a rocket type and a shower type. The nozzle through which the fuel flows is identical in both, with a diameter of 0.9 mm. Unlike a domestic stove where there are two nozzles, for the outer and inner rings, in industrial burners, there is typically only one nozzle.

Puttapoun et al. [20] substituted biomethane for LPG in a ceramic shuttle kiln. The procedure outlined by [11] for domestic stoves was followed. The exit nozzle is modified to keep the velocity of the biomethane similar to the LPG velocity. This allowed entrainment of the correct amount of combustion air. Similar to domestic stoves, too much air resulted in an unstable flame or blow off while too little air resulted in incomplete combustion or flashback.

5.5.2 *Theoretical Considerations*

Similarly to domestic stoves, a higher volume flow rate of biomethane is needed for equivalent heating rates. Simply increasing the gas supply pressure will not work since that increases the nozzle exit velocity, entraining too much combustion air, resulting in blow off at the tip. The diameter of the nozzle must be enlarged to accommodate the flow. The standard LPG nozzle diameter is 0.9 mm. The properties for biomethane and LPG used are shown in Table 5.2.

The exit velocity can be calculated from

$$v_{exit} = \dot{m}_{fuel}/\rho_{fuel}A_{nozzle} \tag{5.13}$$

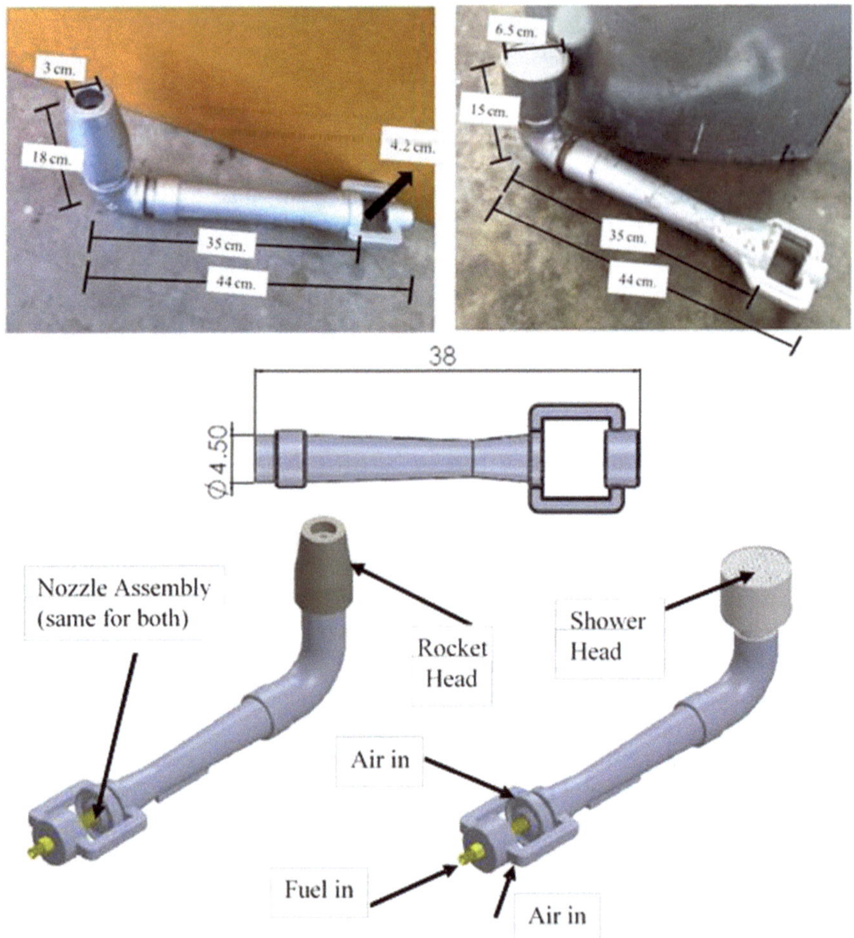

Fig. 5.12 Rocket- and shower-type nozzle heads (Reprinted with permission from [20])

And the output energy is calculated from

$$\dot{Q}_{out} = \dot{m}_{fuel} LHV_{fuel} \tag{5.14}$$

A graph of the fuel energy output versus the exit velocity for five different biomethane nozzle diameters along with the standard LPG nozzle is shown in Fig. 5.13. The maximum energy flow from an LPG nozzle is approximately 7 kW so from this figure it can be seen that the biomethane nozzle diameter should be somewhere between 1.4 and 1.7 mm in order to keep the exit velocity constant.

For an ideal nozzle, the pressure drop between LPG and biomethane can be compared from the Bernoulli formula:

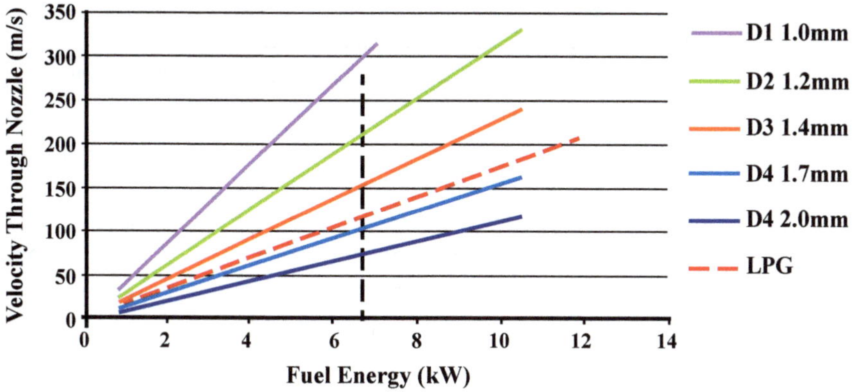

Fig. 5.13 Energy flow through nozzle versus the exit velocity for biomethane and LPG (Reprinted with permission from [20])

$$\Delta P_{fuel} \approx \rho \frac{v^2}{2} \tag{5.15}$$

Substituting from Eq. 5.13 for the exit velocity gives

$$\Delta P_{fuel} \approx \frac{\dot{m}_{fuel}}{\rho_{fuel} A^2} \tag{5.16}$$

Assuming that it is desired to have equivalent heating rates between LPG and biomethane:

$$(\dot{m}LHV)_{LPG} = (\dot{m}LHV)_{biomethane} \tag{5.17}$$

Combining Eqs. 5.16 and 5.17 gives us the comparison:

$$\frac{\Delta P_{biomethane}}{\Delta P_{LPG}} = \left(\frac{LHV_{LPG}}{LHV_{biomethane}}\right)^2 \frac{\rho_{LPG}}{\rho_{biomethane}} = 4.3 \tag{5.18}$$

where the value 4.3 comes from the properties in Table 5.2. According to Eq. 5.18, if the supply pressure for LPG is 6.9 kPa, then the pressure necessary for equivalent heating rates from biomethane would be approximately 30 kPa.

5.5.3 Substitution Results in Kilns

In a shuttle kiln, the injection nozzle was modified in order to substitute biomethane for LPG. The regular LPG nozzle diameter was 0.9 mm and the supply pressure is

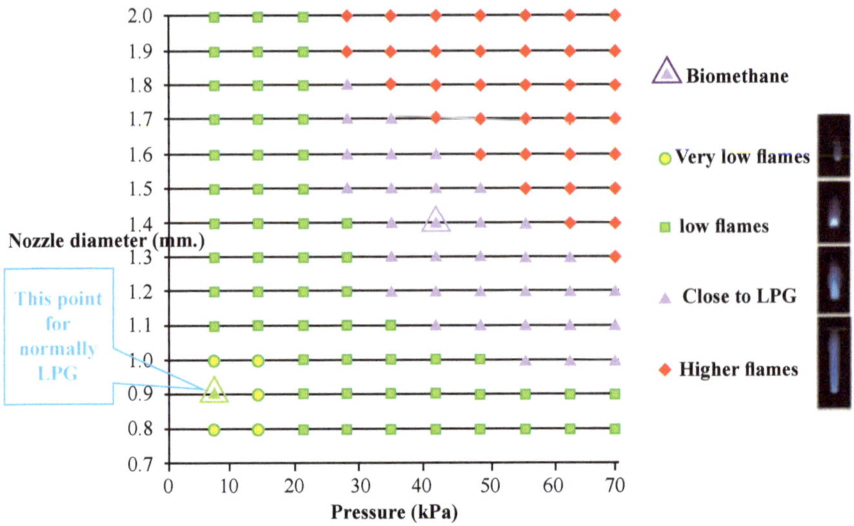

Fig. 5.14 Biomethane substitution results in a shuttle kiln (Reprinted with permission from [20])

6.9 kPa. The nozzle diameter was varied from 1.0 to 2.0 mm in increments of 0.1 mm. The fuel supply pressure was varied between 7 and 70 kPa with the resulting flames sizes monitored. The results from these tests are shown in Fig. 5.14. The circular markers are points where the flame was too high and had a yellow color which results from combustion with insufficient air. The diamond-shaped markers are points where the flame lifted off the burner head, a result of too much entrained air relative to the fuel. The triangular markers represent a stable flame that appeared to match the LPG flame.

As shown in Fig. 5.14, stable flames are at a pressure of 35 kPa and nozzle diameters between 1.2 and 1.8 mm as theoretically predicted from Fig. 5.13 and Eq. 5.18. A mid-range value, a diameter of 1.4 mm, is selected as the optimal nozzle and recommended for use with biomethane. At this diameter and pressure, the average biomethane energy output was 3.1 kW. For comparison, the average output from the LPG nozzle was 2.9 kW. Figure 5.15 shows the biomethane flame at these settings and the LPG flame at its optimal setting (6.9 kPa, 0.9 mm diameter nozzle, 2.9 kW).

In Sect. 5.4, the flame efficiency was discussed. Equation 5.12 was used to determine the efficiency of the biomethane flame which was compared to the LPG for the shuttle kiln. The LPG gave efficiencies from 52.3 to 55.8. The biomethane fuel gave efficiencies between 52.2 and 56.5 which are almost identical in practical terms.

A type R thermocouple traversed laterally across the biomethane flame. From Fig. 5.16, the flame temperatures of both fuels in both the rocket and shower burners are almost identical at all distances from 5 cm from the flames.

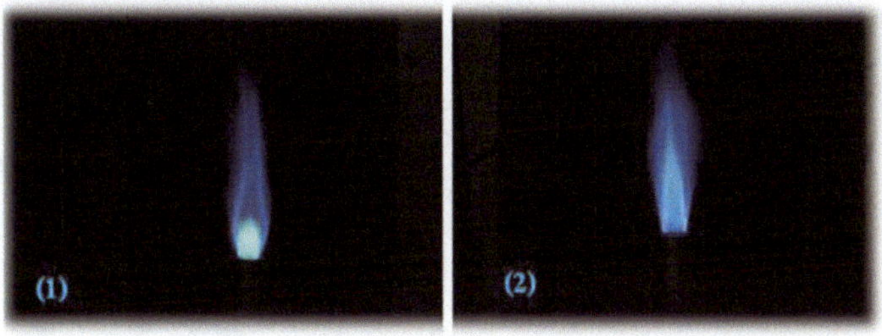

Fig. 5.15 (1) LPG flame, 6.9 kPa, 0.9 mm, 2.9 kW: (2) Biomethane flame, 35 kPa, 1.4 mm, 3.1 kW (Reprinted with permission from [20])

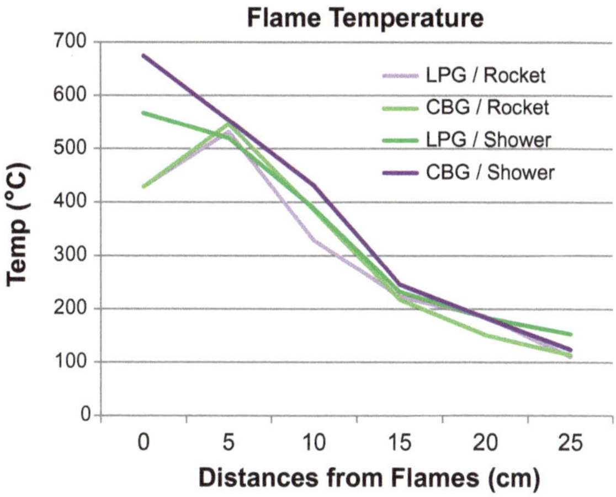

Fig. 5.16 Temperature versus distance for biomethane and LPG flame (Reprinted with permission from [20])

5.5.4 Emission Testing

Emission testings were carried out with a combustion analyzer according to the standard EN 203-1:[21]. The conditions for both tests were at the optimal settings for both fuels as shown in Fig. 5.15.

Figure 5.17 shows the pollutants CO and NOx, corrected to a state of 0% excess oxygen, for both fuels. Data is presented for the rocket- and shower-type burners, respectively. For LPG, the measured CO was 3586 and 3368 ppm for which the corresponding levels for biomethane were 1062 and 2019 ppm. This represents an average drop of 44%. Similarly, the NOx levels of LPG were 24.7 and 22.9 ppm for both burners. For the biomethane, the levels were 6.5 and 14.7 ppm which also have an average drop of 44%.

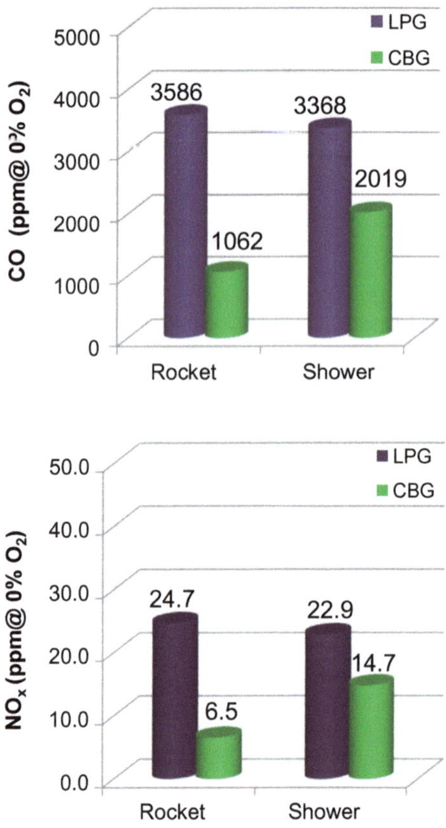

Fig. 5.17 Emissions of CO and NOx from LPG and biomethane (Reprinted with permission from [20])

5.6 Economics of Fuel Substitution

The Thai government subsidizies the use of LPG, although for industrial use, LPG is subsidized less than for domestic cooking and transportation. LPG price has been controlled by the government since gas was discovered in the Gulf of Thailand in 1981 [22]. The industrial price of LPG over the past 8 years in US dollars is shown in Fig. 5.18. The price rose in 2012–2014 due to a reduction in the subsidy and it fell a little in 2016 in response to falling petrochemical prices internationally. A more complete picture of the Thailand's energy policies can be found at [23].

Industrial substitution with biomethane would benefit the local economy as well as cost savings. The changeover from LPG to biomethane has an economic cost. The equipment cost incurred for using LPG is shown in Table 5.5 and the cost for installing a biomethane fuel delivery system is shown in Table 5.6. This biomethane delivery system has been described before in [24]. The biomethane is stored in 100 L tanks at a pressure of 20MPa as shown in Fig. 5.19. The biomethane system costs an extra $1,765 (฿58,323) to install. The operators need to be able to recoup this

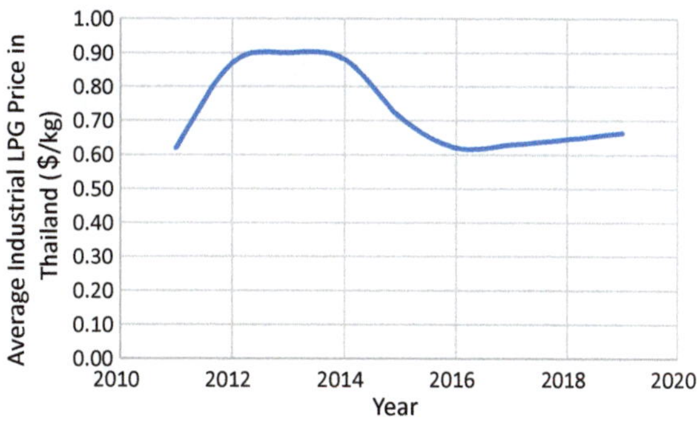

Fig. 5.18 Price of LPG for industrial use in Thailand

Table 5.5 Capital costs for installing LPG (Reprinted with permission from [20])

Cost of LPG			
Number	Equipment	Quantity	Price ($)
1	Tank of LPG 48 L tank of LPG	1	86.75
2	Low-pressure regulator	1	11.4
Total			98.15

Table 5.6 Capital costs for installing a CBG system (Reprinted with permission from [20])

Cost of compressed biomethane (CBG)		
Number	Equipment	Price ($)
1	CNG-Type 1 100 L Tank	1091
2	Iron frame structure	212
3	High-pressure regulator	71.4
4	Gauges, piping, valves nozzles, and fittings	491
Total		1865.4

cost from savings on the running costs. The cost of CBG depends on the production capacity of the raw biogas as shown in Fig. 5.20. Without redoing the analysis from Sect. 3.2, the production costs are divided into five categories:

1. Cost of producing biogas,
2. Capital cost of the upgrading unit,
3. Operating and maintenance costs of the upgrading unit,
4. Capital cost of the compressor, and
5. Operating costs of the compressor.

These five costs are dependent on the quantity of biomethane produced. The capital and operating costs per kg of biomethane produced decrease with quantity of biomethane. So, for example, if the raw biogas is produced at a rate of

Fig. 5.19 Two 100 L tanks tanks of CBG (Type 1) at 20 MPa (Reprinted with permission from [20])

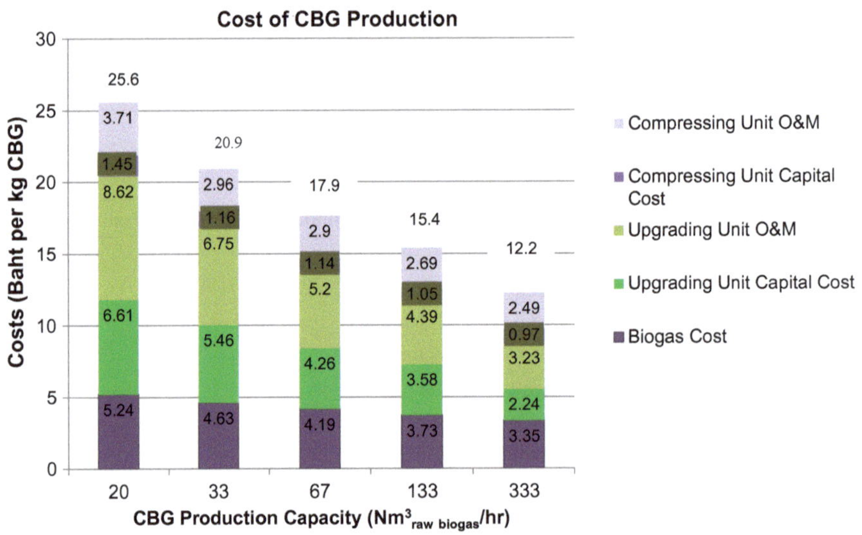

Fig. 5.20 Production costs of CBG as a function of production capacity from [25]

$20\,\text{Nm}^3_{raw\,biogas}/\text{h}$, the cost of the compressed biomethane would be $\$0.77/\text{kg}_{CBG}$ ($\$17.17/\text{GJ}_{Biomethane}$) but this drops to $\$0.37/\text{kg}_{CBG}$ ($\$8.25/\text{GJ}_{Biomethane}$) if the production rate jumps to is $333\,\text{Nm}^3_{raw\,biogas}/\text{h}$ [26]. Taking an industrial LPG price in 2019 of $\$0.70/\text{kg}_{LPG}$ ($\$15.21/\text{GJ}_{LPG}$) shows the importance of having an economic production process for biomethane. At low production rates, it is not competitive but at high rates (>300 Nm^3/h) it becomes cost-effective to switch fuel.

5.7 Case Study—Mee Sin Ceramics

Modifying the burner to run on biomethane appeared to give positive results. The flame visually looked the same as an LPG flame, the temperature profiles were similar, similar efficiencies and the pollutants were lower. However, it cannot be stated with certainty that biomethane can be used as an LPG replacement until it is tested and operated under normal industrial conditions. Biomethane was fired in a ceramic kiln and the products compared with ceramic products from LPG firing. These tests were performed with a 0.1 m^3 shuttle kiln in Lampang, a province in the north of Thailand, at the Mee Sin Ceramics company, see Fig. 5.21.

The ceramic kiln used was a 0.1 m^3 volume kiln as shown in Fig. 5.21. A kiln volume refers to the volume of product inside it and not the actual inner volume. It is heated via four burner heads. At the top of the kiln in the center is a thermocouple. The exhaust gases exit through a center port and out through a stack. A damper is located in the stack. By closing the damper, the pressure and temperature inside the kiln increase. Another method to increase the temperature is to increase the fuel firing rate.

Twenty bowls of unfired clay articles, or greenware, were fired each time. The combined weight was between 11 and 12 kg, as shown in Fig. 5.21. The firing process is seven and a half hours long during which the temperature is ramped at a rate of 2 °C per minute up to a final temperature of 1200 °C. The temperature graph is shown in Fig. 5.22. Superimposed on this theoretical graph are the actual temperature measurements from firing with both LPG and biomethane.

The ceramic products are shown in Fig. 5.22. Biomethane is capable of accurately tracking the required temperature rate. Once the ceramic bowls had cooled they were removed from the kiln and inspected for cracks, chipping, damaged paintwork, water

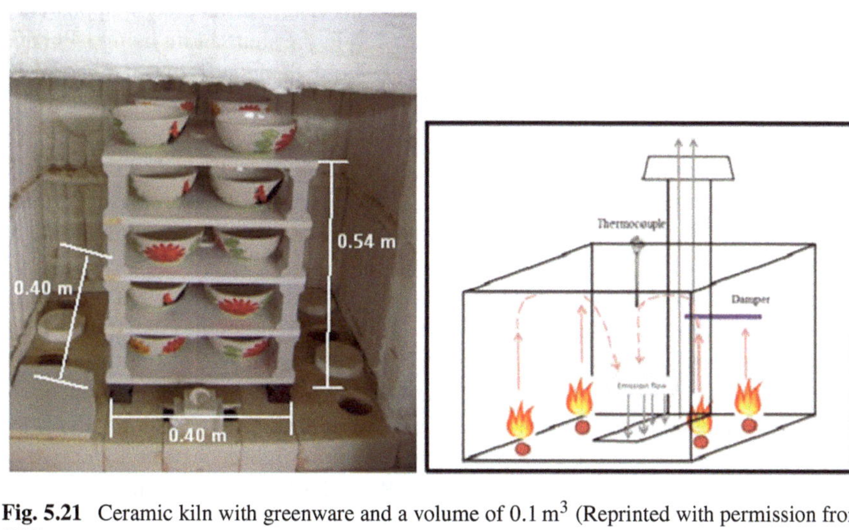

Fig. 5.21 Ceramic kiln with greenware and a volume of 0.1 m^3 (Reprinted with permission from [20])

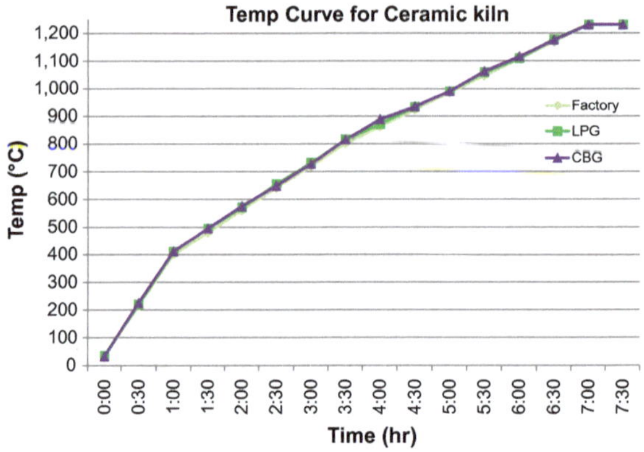

Fig. 5.22 Temperature curve for firing greenware (Reprinted with permission from [20])

Fig. 5.23 Comparison of final products fired with LPG and biomethane (Reprinted with permission from [20])

leaks, and any other sign of defects. There was no visible difference in appearance between the factory default bowls and those fired in the shuttle kiln using LPG and biomethane. None of the 60 bowls fired with biomethane suffered damage to its surface or color as can be seen in Fig. 5.23.

Table 5.7 displays the costs of firing with this 0.1 m^3 shuttle kiln. Each firing lasts approximately seven and a half hours and uses 16.5 kg of LPG or 22.4 kg of CBG. With an annual total of 144 firings per year yielded annual fuel savings of around \$374.7 (฿12,365). The CBG price was assumed to come from an economic biogas plant.

The economics of changing to biomethane are displayed in Table 5.8. It assumes a 10-year life span and an interest rate of 7.0%. The capital cost of the CBG comes from Table 5.6 and the cost of CBG from Fig. 5.20. The internal rate of return is 17.0% and the payback period is 4 years and 8 months. This analysis is highly dependent on the LPG and compressed biomethane gas price. A slight increase in LPG price resulting from removing the Government subsidy or a slight reduction in the cost of biomethane leads to a large increase in the IRR and NPV. Slight price movements in the opposite direction leads to a decrease in IRR and NPV.

Table 5.7 Annual fuel savings for a CBG system (Reprinted with permission from [20])

Description	Details
Ceramic Kiln	0.1 m^3
Firing times per year	144*
Kg of LPG used in one firing (kg./1 time)	16.50
Kg of CBG used in one firing (kg./1 time)	22.40
Price of LPG ($/kg.)	$0.66**
Price of CBG ($/ kg.)	$0.37
Fuel savings per firing ($/time)	$2.60
Fuel saving per annum ($/year)	$374.7

*One day firing, one day cooling for a total of 288 days per year
**Thai industrial price of LPG in June 2019

Table 5.8 Economic analysis of changing over to CBG in a shuttle kiln (Reprinted with permission from [20])

Description	Details
Cost for changing LPG to CBG	$1767.4
Life for CBG tank (Years)	10
Interest rate assumed for analysis	7.0%
Fuel saving ($/year) from Table 5.7	$374.7
NPV	$808.0
IRR	17.0%
Payback period (Years)	4.7

5.7.1 Mee Sin Ceramics Discussion

The purpose of this chapter was to introduce the possibility of substituting biomethane for LPG in domestic and industrial stoves. Environmental and economic reasons are the primary motivation for this switch. The biomethane used in these studies contained a minimum of 85% pure methane. In an industrial-sized burner, to switch over from LPG to biomethane has a payback period of roughly 4 years and 8 months. This economic performance should improve as the price of biomethane continues to fall as economies of scale are developed and the price of LPG remains the same or increases due to a global increase in the hydrocarbon price. The payback period fluctuates as the price of LPG and/or biomethane fluctuates. The LPG nozzle needs an increase in size in order to give a clean stable flame. For an LPG nozzle diameter of 0.9 mm and an LPG supply pressure of 6.9 kPa, the equivalent biomethane nozzle diameter is 1.4 mm and its fuel supply pressure is 34.5 kPa. The temperature profile and efficiency results are found to be very similar for both fuels. The emissions of CO and NO_x were found to be approximately 44% lower for the biomethane flame. Finally, a case test was performed at the Mee Sin Ceramics at Lampang with no discernible difference between the ceramic bowls fired using biomethane. The conclusion is that it is possible to change from LPG to biomethane for use in the ceramic industry.

References

1. EPPO (2018) Energy statistics. http://www.eppo.go.th/index.php/en/en-energystatistics/petroleumprice-statistic?orders[publishUp]=publishUp&issearch=1
2. Prasertsant P, Sajjakulnukit B (2006) Biomass and biogas energy in thailand: potential, opportunity and barriers. Renew Energy 31:599–610
3. Yokoyama S, Ogi T, Nalampoon A (2000) Biomass energy potential in Thailand. Biomass Bioenergy 18:405–410
4. Chaiprasert P (2011) Biogas production from agricultural wastes in thailand. J Sustain Energy Environ 63–65
5. Aggarangsi P, Tippayawong N, Moran J, Rerkkriangkrai P (2013) Overview of livestock biogas technology development and implementation in thailand. Energy Sustain Dev 17:371–377
6. Nasir IM, Ghazi T, Omar R (2012) Anaerobic digestion technology in livestock manure treatment for biogas production: a review. Eng Life Sci 12:258–269
7. Sakar S, Yetilmezsoy K, Kokac E (2009) Anaerobic digestion technology in poultry and livestock waste treatment for biogas production: a literature review. Waste Manag Res 27:3–18
8. Gunaseelan VN (1997) Anerobic digestion of biomass for methane production: a review. Biomass Bioenergy 13:83–114
9. Bond T, Templeton MR (2011) History and future of domestic biogas plants in the developing world. Energy Sustain Dev 15(4):347–354
10. Dai W, Qin C, Chen Z, Tong C, Liu P (2012) Experimental studies of flame stability limits of biogas flame. Energy Convers Manag 63:157–161
11. Suwansri S, Moran J, Aggarangsi P, Tippayawong N, Bunkham A, Rerkkriangkrai P (2015) Converting lpg stoves to use biomethane. Energy Sustain Dev 30(1):38–57
12. Tanatvanit S (1998) The relationship between performance and emission of LPG cooking stove. Master's thesis, King Mongkuts University of Technology, Thonburi
13. Jugjai S, Tia S, Trewetasksorn W (2001) Thermal efficiency improvement of an lpg gas cooker by a swirling central flame. Int J Energy Res 25:657–674
14. Jugjai S, Tia S, Tia V, Thaneswanich S (2007) Performance testing of LPG cookstoves in Thailand. Technical Report, Energy Policy and Planning Office
15. Lucky RA, Hossain I (2001) Efficiency study of bangladeshi cookstoves with an emphasis on gas cookstoves. Energy 26:221–237
16. Razus D, Oancea D, Brinzea V, Mitu M, Munteanu V (2007) Experimental and computational study of flame propagation in propane-n-butane and liquefied petroleum gas-air mixtures. In: 3rd European combustion meeting, Chania, Greece
17. Anggono W, Wardana I, Lawes M, Hughes K, Wahyudi S, Hamidi N, Hayakawa A (2013) Biogas laminar burning velocity and flammability characteristics in spark ignited premix combustion journal of physics conference series 423. J Phys Conf Ser 1–7
18. Mazas N, Fiorina B, Lacoste D, Schuller T (2011) Effects of water vapor addition on the laminar burning velocity of oxygen-enriched methane flames. Combust Flame 158:2428–2440
19. Yan B, Wu Y, Liu C, Yu J, Li B, Li Z, Chen G, Bai X, Alden M, Konnov A (2011) Experimental and modeling study of laminar burning velocities of biomass derived gas/air mixtures. Int J Hydrog Energy 36:3769–3777
20. Puttapoun W, Moran JC, Aggarangsi P, Bunkham A (2015) Powering shuttle kilns with compressed biomethane gas for the thai ceramic industry. Energy Sustain Dev 28:95–101

21. European Committee for Standardization (2014) En 203-1:2014. gGas heated catering equipment. General safety rules. Technical report, European Standard
22. Office of Energy Policy and Planning (2011) Report on policies of the LPG price structure. http://www.escctcc.com/upload/Page/default_knowledge_information/general_lpg.pdf
23. Energypedia (2018) Fuel prices Thailand. https://energypedia.info/wiki/Fuel_Prices_Thailand. Accessed 12 June 2018
24. Koonaphapdeelert S, Kanta U, Aggarangsi P (2011) Biomethane: an alternative green fuel to CNG. In: 7th international conference on automotive engineering, Bangkok
25. Koonaphapdeelert S, Kanta U, Aggarangsi P (2011) Biomethane: an alternative green fuel to CNG. In: 7th international conference on automotive engineering, Bangkok
26. ERDI (2013) A prototype bio-methane gas compressor for automotive applications (in thai). Technical report, Energy Research and Development Institute, Chiang Mai
27. Suwansri S, Moran J, Aggarangsi P, Tippayawong N, Bunkham A, Rerkkriangkrai P (2014) Converting lpg stoves to biomethane. Distrib Gener Altern Energy 29(4)

Chapter 6
Biomethane to Local Gas Grids

6.1 Biomethane to Local Gas Grids

In situations where it is not feasible to inject biomethane into a national methane network because of regulation or unavailability, an alternative may be to build a small local gas grid. The biomethane can then be piped to the local community. It can be used for cooking and heating in local homes and businesses as described in Chap. 5.

Since an international definition for a "local biomethane grid" is not available, the following shall be taken as a suitable definition for the time being. "A biomethane grid is a network for the distribution of locally produced biogas/biomethane to end users in a local town or city to meet demand for a cleaner more efficient environmentally friendly energy source." This chapter will outline a methodology for designing, building, and operating a biomethane grid. There are requirements and procedures necessary to follow before installing and operating a biomethane grid.

A good starting point in designing a biomethane grid is the American Society of Mechanical Engineers standard B31.8 (ASME [1]) which has been incorporated into American federal law. This is a large and complicated standard which deals with all aspects of natural gas grids. This standard is suitable for building cross-continental pipelines from massive gas fields. For a small, local gas grid, this standard is overkill. Many aspects are unnecessary. A more realistic standard was developed by Moran et al. [2] for local biomethane grids in Thailand. Some of the basics of this standard are summarized below:

1. The methane CH_4 content shall be at least 85%. The contents of the remaining 15% is explained in Sect. 6.1.1.
2. The pressure in the biomethane grid shall not exceed a gauge pressure of 400 kPa. This is because of the small grid size a large pressure is not required. Also, it means that the only pressure-limiting device needed is a regulator located at the customer residence capable of reducing distribution line pressure to pressures recommended for household appliances.
3. The hoop stress on the pipeline shall be less than 20% of the specified minimum yield strength of the pipe.

S. Koonaphapdeelert et al., *Biomethane*, Green Energy and Technology,
https://doi.org/10.1007/978-981-13-8307-6_6

4. If the biomethane grid consists of metal piping it shall operate between a temperature range from −10 to 230 °C only. For plastic piping, the temperature range depends on the plastic.
5. No other gases shall be transported except biogas or biomethane.
6. The standard developed only applies to the external use of biomethane and its transport across public lands. It does not apply when the gas is used on site for private purposes.

6.1.1 Biomethane Purity

There is no international standard for biomethane purity particularly for natural gas grid injection. Standards do exist for processed and blended natural gas [3] but not biomethane. Biomethane purity standards are set by individual countries. For example, the Netherlands requires at least 85% of CH_4 content in the injected biomethane, while in Switzerland and Sweden, the requirements are for 96% and 97% CH_4, respectively [4]. The biomethane should be set to a sufficiently high purity level so that it can be safely combusted at the customer's end location.

The European Committee for Standardization (CEN) is a public standards organization which has attempted to create a European wide standard for biomethane injection into gas grids, European Committee for Standardization [5]. The standard is for H and L gas networks. These are high- and low-calorific gas grades. Parts of the European gas market are supplied with low-calorific natural gas (L-gas) which is exclusively sourced from German and Dutch reserves. All other supplies from Denmark, Norway/North Sea, Russia, or via LNG terminals are high-calorific gas (H-gas). For technical reasons and due to weights and measures regulations, the two gases have to be transported in separate systems. Customers receiving one type of gas cannot switch over to gas of the other quality unless their gas appliances are adjusted beforehand, similar to the fuel switching issues outlined in Chap. 5. The biomethane standard is outlined in Table 6.1. However, this standard only applies to certain impurities. Other biomethane properties such as its calorific value, methane quantity, Wobbe index, relative density, and CO_2 content are left as requirements for national governments to specify.

At the national level, the purity of the biomethane is set to a level where it can be injected into the national gas grid. Table 6.2 shows the standards for the methane content in biomethane across seven European countries. Once this level is reached the biomethane can legally be injected into the respective national gas grids. As can be seen, all have a minimum methane content above 85%.

Holland has one of the lowest standards for biomethane purity levels since its natural gas grid is designed around the gas purity level of their offshore Groningan gas field [7]. Table 6.3 shows the national specifications for injecting biomethane into the high-pressure distribution network of the Dutch gas grid. The Singaporean natural gas standard is shown in Table 6.4.

Table 6.1 Applicable common requirements and test methods for biomethane at the point of entry into high-calorific gas and low-calorific natural gas networks

Parameter	Unit	Limit values		Test method (Informative)
		Min	Max	
Total volatile silicon (as Si)	mgSi/m^3		0.3–1	EN ISO 16017-1:2000 TDS-GC-MS
Compressor oil		The biomethane shall be free from impurities other than "de minimis" levels of compressor oil and dust impurities		ISO 8573-2:2007
Dust impurities		The biomethane shall be free from impurities other than "de minimis" levels of compressor oil and dust impurities		ISO 8573-4:2001
Chlorinated compounds			See CEN/TR, (WI 00408007)	EN 1911:2010
Fluorinated compounds				NF X43–304:2007 ISO 15713:2006
CO	% mol		0.1	EN ISO 6974-series
NH_3	mg/m^3		10	NEN 2826:1999 or VDI 3496 Blatt 1:1982–04 NF X43–303:2011
Amine	mg/m^3		10	VDI 2467 Blatt 2:1991–08

Table 6.2 Minimum methane content in biomethane for some European Countries [6]

Component	Austria	France	Belgium	Czech R
CH_4	≥96%	≥86%	≥85%	≥95%
	Holland	Sweden	Switzerland	
CH_4	≥85%	≥97%	≥96%	

For a local biomethane gas grid, the biomethane purity level can be set at a level that satisfies the local community. There are two factors that suggest a high purity >85% may be preferable. The first concerns the Wobbe Index (WI) which is an indicator of the interchangeability of fuel gases such as methane, Liquefied Petroleum Gas (LPG), and different grades of biogas. It is frequently defined in the specifications of gas supply and transport utilities and has been discussed in Sect. 5.4.1. As explained,

Table 6.3 Properties of biomethane for grid injection (high pressure, L-value) in the Netherlands [8]

Component	Fraction	Component	Fraction	Component	Fraction
CH_4	≥85% (vol/mol)	Stot	≤5.5 (mgS/m^3)	Siloxanes	<0.1 as Si (mg/m^3)
CO_2	≤3% (mol)	$H_2S + COS$	≤5 (mgS/m^3)	Ammonia H_2O	≤3 (mg/Nm^3)
O_2	≤0.0005% (mol)	Mercaptan sulfur	≤6 (mgS/m^3)	Water dew point (°C at 70 bar abs.)	≤ − 8
H_2	≤0.2% (mol)	Halocarbons	≤5 (mg/m^3) (Cl/F)	Wobbe index	43.46–44.41 MJ/m^3
CO	≤2900 (mg/m^3)	Heavy metals	N/A	Dust	≤100 Size >5 μm

Table 6.4 Characteristics of natural gas for use in Singapore [9]

Component	Fraction	Component	Fraction	Component	Fraction
CH_4	≥80% (vol/mol)	S tot	≤30 ppm	Magnesium	≤2 ppm by weight
CO_2	≤5% (vol/mol)	H_2S	≤8 ppm	Ammonia H_2O	≤3 (mg/Nm^3)
O_2	≤0.1% (vol/mol)	Potassium and Sodium	≤5 ppm by weight	Water dew point	9.4 °C @ 50 bar
CO	N/A	Lead	≤1 ppm by weight	Particles	<3 ppm by weight

gases of similar Wobbe index can be used in the same equipment. So, for example, a natural gas cooker or heater designed for gas with a WI of 50 MJ/m^3 could not be used with a gas having a WI of 23 MJ/m^3. An example of a gas with a WI of 23 MJ/m^3 is biogas with a methane percentage of 60%. So any equipment that combusts 60% biogas needs to be specially modified to burn this fuel, usually by enlarging the fuel nozzle [10]. This poses a safety risk for customers if they connect methane or LPG burners to a 60% biogas pipeline. The gas will flow out but not ignite causing a dangerous scenario.

The exact range of Wobbe Index that accommodates stable and safe combustion is largely dependent on the application and equipment. To be completely certain, it is necessary to test the gas with the appliance. The European Association for the Streamlining of Energy Exchange (EASEE) has proposed a higher Wobbe Index range from 48.96 to 56.92 MJ/m^3 [11], to streamline pipeline interoperability at cross-border points in Europe. This corresponds to a gas relative density between 0.55 and 0.7 which is in the range of densities for biomethane as the methane content varies from 100 to 85%. The range is quite broad and some engine manufacturers have raised concerns that this range may not be ideally suited to their engines [12].

Table 6.5 Advantages and disadvantages of using differing biomethane purity levels in local gas grids

	Advantages	Disadvantages
Biomethane containing at least 85% methane	• Meets minimum standard for European countries • Can use existing natural gas burners/heaters [14] • Can be used in NGVs as a substitute for natural gas [13] • Heating value of 30.4 MJ/m^3 which is comparable to pure methane which is 36.14 MJ/m^3	• Higher processing cost • Higher initial investment
Biomethane containing at least 60% methane	• Lower processing cost • Lower capital investment	• Must have specially modified burners at point of end use. Potential safety issue if improperly used at customer site • Cannot be used in NGVs • Low heating value of only 21.48 MJ/m^3 and a longer cooking time • Lower than most international gas grid standards

Another issue is the heating value of the biomethane. For similar sized flames, a low heating value means more time is needed to heat the end product. This may dissatisfy some customers. This suggests it is better to have a high percentage of methane, above 85%, in the final product that gets used in a biomethane grid. Processing biogas into biomethane containing at least 85% methane would eliminate both of the above issues and also be of sufficient quality for use as a natural gas substitute in vehicles [13]. Methane content of 85% is also the proposed minimum for NGVs in Thailand. These issues are presented in Table 6.5.

The standard for a biomethane grid developed by Moran et al. [2] recommends using a biomethane mixture containing a minimum of 85% methane. The advantages, especially the safety aspects were thought to outweigh the disadvantages. So long as the biomethane in the local grid is not being used in specialized engines, then the range of 85% is deemed sufficiently broad to satisfy both consumers and biogas producers. The final list of the biomethane components and the limits are shown in Table 6.6.

6.1.2 Existing Pipeline Standards

There are many standards that exist for gas pipelines and operations associated with them. The following is a brief description of some of the important and relevant

Table 6.6 Proposed list of components and their limits proposed in the local grid standard

Components	Limits	Explanation
CH_4	≤85%	Ease of use. No need to modify combustion equipment differently
CO_2	<18% vol	Corrosion protection
O_2	<1% vol	Protection from accidental ignition
H_2	–	Normally very little in a biogas system
CO	–	Normally very little in a biogas system
S tot	≤45 mg/Nm^3	Similar to Thai NGV/ Holland
H_2S	≤10 mg/Nm^3	Slightly higher criteria of safety
Mercaptanes*	–	Already included in Stot
Halocarbons*	–	Normally very little in the system
Heavy metals*	–	Normally very little Hg in the system
Siloaxanes*	–	More likely to be in landfill gas than gas from agricultural waste
Ammonia. H_2O	≤20 mg/Nm^3	Obtained from levels allowed in Sweden and Switzerland
Water dew point	≤−10 °C	From Holland and Thai NGV standard
Odorant	Enough to be detectable by a normal person at 20% of gas flammability limit	From ISO 13734
Particles	No particle	A level specified in Germany

*If biogas comes from a landfill site these components must follow standard international norms

standards that may be useful for use in local gas grids. A commonly used standard in this field is the American Society of Mechanical Engineers standard B31.8 [1]. This is an American standard for Gas Transmission and Distribution Piping Systems. ASME standards are accepted for use in more than 100 countries around the world. This is partly because they are developed in a committee setting to ensure balanced participation and open access to public interest groups. ASME uses a consensus process, where meetings dealing with standards-related actions are open to all members of the public. This provides us with confidence that if a recommendation is made based on ASME B31.8 it has been rigorously analyzed and agreed upon by subject matter experts.

ASME B31.8 lists out acceptable materials from different grades of steel, iron, thermoplastics, and epoxy resins that are acceptable materials for use in a gas piping network. It was decided to use the same material standards in the local biomethane gas grid. Any contractor constructing a biomethane local grid should investigate the specific pipe, tubing, or fitting to be used and should determine material serviceability for the conditions anticipated. Whatever material is selected should adequately resist liquid and chemical atmospheres that may be encountered. Different materials getting joined together should be ensured beforehand that they are compatible with each other. The various valves, flanges, and fittings that are part of a gas grid should also adhere to the standard specified in ASME B31.8. It is important that each component of a pipeline must be able to withstand the operating pressures and other anticipated loadings without damage or impairment of its serviceability. It is a good idea that any pipe or components in which liquids may accumulate have drains or drips. Another good practice is for pipes or components subject to clogging from solids or deposits to have suitable connections for cleaning.

6.1.3 Class Locations

The factor of safety used in pipeline should depend on the pipe location. A gas pipeline going through the center of New York should have a higher factor of safety than a pipeline in the middle of the countryside. To deal with this issue, the concept of a class location is introduced. This is based on the ASME B31.8 standard. A "class location unit" is defined as an area that extends 200 m on either side of the pipeline centerline of any continuous 1.6 km length of pipeline, see Fig. 6.1. Even though Fig. 6.1 shows the pipeline as a straight line, this does not have to be the case. A unit location tracks any 1.6 km length of pipeline and extends 200 m to either side of it even if the pipeline has many curves and bends.

1. A Class 1 location is: Any class location unit that has 10 or fewer dwellings intended for human occupancy.
2. A Class 2 location is: Any class location unit that has more than 10 but fewer than 46 dwellings intended for human occupancy.
3. A Class 3 location is: (i) Any class location unit that has 46 or more dwellings intended for human occupancy or (ii) an area where the pipeline lies within 91 m of either a building or a small, well-defined outside area such as a playground, recreation area, outdoor theater, or other place of public assembly. This area is occupied by 20 or more people on at least 5 days a week for 10 weeks in any 12-month period. (The days and weeks need not be consecutive.) In other words, if the pipeline is within 91 m of a regular school it counts as a class 3 location.
4. A Class 4 location is: Any class location unit where dwellings with four or more stories above ground are prevalent.

Each separate apartment or room in a multiple apartment building is counted as a separate dwelling in this definition. The border between two class locations maybe taken as follows: (a) A Class 4 location ends 200 m from the nearest building with

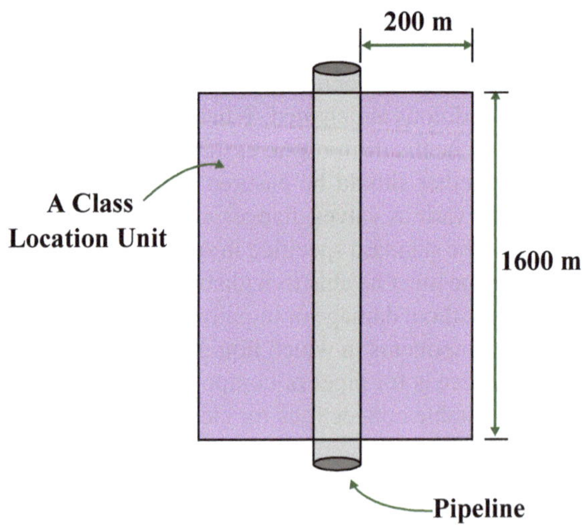

Fig. 6.1 Area of a single class location unit

Table 6.7 Design factors for different class locations (Reprinted with permission from [2]

Class location	Design factor (F)
1	0.72
2	0.60
3	0.50
4	0.40

four or more stories above ground. (b) For a Class 2 or 3 location, the class location ends 200 m from the nearest building in the cluster. The significance of the different classes come into being when designing and testing the pipeline that goes through each class. The table of safety or design factors, F, for the different class locations is shown in Table 6.7.

Except for class 4 locations, the design factor increases a single level for situations where the pipeline crosses a hard surfaced road, a highway, a public street, or a railroad.

6.1.4 Pipe Design

6.1.4.1 Steel Piping

The following formula can be used to determine the maximum inner diameter of the pipeline made from steel [2].

$$D_{max} = 2\left(\frac{S.t}{P}\right) x F x E_f x T \tag{6.1}$$

Table 6.8 Longitudinal joint factor (E_f) for steel pipe

Specification	Pipe class	Longitudinal joint factor (E_f)
ASTM A 53/A53M or equivalent	Furnace butt welded	0.60
API 5 L or equivalent	Furnace butt welded	0.60
Undetermined	Pipe 102 m diameter or more	0.80
Undetermined	Pipe 102 m diameter or less	0.60

Table 6.9 Temperature derating factor (T) for steel pipe (Reprinted with permission from [2])

Gas temperature in degrees Celsius (°C)	Temperature derating factor (T)
120 °C or less	1.00
150 °C	0.967
175 °C	0.933
205 °C	0.90
230 °C	0.867

where

D_{max} = Nominal outside maximum diameter of the steel pipe in millimeters
P = Design pressure in kPa gauge, where the maximum pressure is limited to 400 kPa
S = Yield strength in kPa
t = Nominal wall thickness of the pipe in millimeters (mm)
F = Design factor determined from Table 6.7
E_f = Longitudinal joint factor, usually 1.0, except in cases shown in Table 6.8
T = Temperature derating factor as shown in Table 6.9

The longitudinal joint factor (E_f) to be used in the design formula in Eq. 6.1 is always 1.0 except for the material classes listed in Table 6.8. If the type of longitudinal joint cannot be determined then it is classified as "undetermined".

The temperature derating factor, (T) to be used in the design formula in Eq. 6.1 is determined from Table 6.9.

6.1.4.2 Plastic Piping

For plastic piping, it is also recommended that the pressure is kept below 400 kPa. The maximum diameter for plastic pipe can be determined from the following formula [2]:

$$D_{max} = t + 2S\frac{t}{P}(0.32) \tag{6.2}$$

where

Table 6.10 Selected long-term hydrostatic strength (S) for certain thermoplastic materials

Plastic pipe material designated by ASTM D 2513	Long-term hydrostatic strength at 23 °C, kPa
PB 2110	13,500
PE 2406	8,500
PE 3408	11,000
PVC 1120	27,500
PVC 1220	27,500
PVC 2110	13,700
PVC 2116	21,500

Table 6.11 Minimum wall thicknesses for reinforced thermosetting plastic pipe

Nominal size in millimeters	Minimum wall thickness millimeters
50	1.6
75	1.6
100	1.8
150	2.5

D_{max} = Nominal outside maximum diameter of the pipe in millimeters
P = Design pressure in kPa gauge, where the maximum pressure is limited to 400 kPa
S = For thermoplastic pipe, the long-term hydrostatic strength (HDB) in accordance with ASTM D 2513 at a temperature equal to 23 °C, see Table 6.10. For reinforced thermosetting plastic pipe, use 75,500 kPa
t = Nominal wall thickness of the pipe in millimeters. For thermoplastic pipe, this thickness should always be greater than 1.6 mm

The temperature limits for operation using any plastic pipe are from −25 to 60 °C. For thermoplastic pipe, if the outside diameter is greater than 22 mm then the wall thickness must be at least 2.3 mm. The wall thickness for reinforced thermosetting plastic pipe may not be less than that listed in Table 6.11:

6.1.4.3 Hoop Stress

The Hoop Stress on a pipe is given by Barlow's formula:

$$S_H = \frac{PD}{2t} \tag{6.3}$$

where

S_H = Hoop stress on pipe (kPa).
P = Maximum allowed pressure in the pipeline (400 kPa).

Table 6.12 Minimum cover requirements for biomethane grid pipeline

Location	Cover required (mm)
Class 1 locations	800
Class 2, 3, and 4 locations	1,000
Drainage ditches of public roads and railroad crossings	1200
Waterways, lakes, rivers, etc.	1,500

D = Pipe diameter in millimeters.
t = Pipe thickness in millimeters.

It is recommended that the maximum hoop stress be kept below 20% of the specified minimum yield strength of the pipe material. Having a hoop stress this low has several advantages including a large factor of safety, any welds need only to be visually inspected, orange-peel bull plugs and orange-peel swages are allowed, less laborious inspection procedures, and less onerous testing requirements.

6.1.4.4 Flexibility

Each pipeline must be designed with enough flexibility to prevent thermal expansion or contraction from causing excessive stresses in the pipe or components, excessive bending or unusual loads at joints, or undesirable forces or moments at points of connection to equipment, or at anchorage or guide points.

6.1.5 General Construction Requirements

6.1.5.1 Underground Clearance

Once again for a local biomethane gas grid, each underground pipeline must be installed with at least 305 mm of clearance from any other underground structure not associated with the pipeline. If this clearance cannot be attained, the transmission line must be protected from damage that might result from the proximity of the other structure. Each pipeline must be installed with enough clearance from any other underground structure to allow proper maintenance and to protect against damage that might result from proximity to other structures.

Table 6.12 gives minimum cover requirements for buried pipelines. Where an underground structure prevents the installation of a pipeline with the minimum cover, the line may be installed with less cover if it is provided with additional protection to withstand anticipated external loads.

6.1.5.2 Protection from Hazards

The operator must take all practicable steps to protect each line in the biomethane grid from washouts, floods, unstable soil, landslides, or other hazards that may cause the pipeline to move or to sustain abnormal loads. When a ditch for a pipeline is backfilled, it must be backfilled in a manner that provides firm support under the pipe and prevents damage to the pipe and pipe coating from equipment or from the backfill material. The soil should be returned to the trench with the subsoil put back first, followed by the topsoil. In areas where the ground is rocky and coarse, the workmen will either screen the backfill material to remove rocks, bring in clean fill to cover the pipe, or cover the pipe with a material to protect it from sharp rocks. Once the pipe is sufficiently covered, the coarser soil and rock can be used to complete the backfill. Each above ground pipe in the biomethane grid must be protected from accidental damage by traffic or other similar causes, either by being placed at a safe distance from the traffic or by installing barricades.

6.1.5.3 Installation of Plastic Pipe

Plastic pipe must be installed below ground level. If it has to be installed above ground, the operator must demonstrate that the cumulative above ground exposure of the pipe does not exceed the manufacturer's recommended maximum period of exposure or 2 years, whichever is less. The pipeline must also be located where damage by external forces is unlikely or is otherwise protected against such damage. The pipeline must adequately resist exposure to ultraviolet light and high and low temperatures. Plastic pipe that is not encased must have a means of locating the pipe while it is underground. Tracer wire or other metallic elements installed for pipe locating purposes must be resistant to corrosion damage, either by use of coated copper wire or by other means.

6.1.6 Customer Meters and Regulators

Each meter and service regulator, whether inside or outside a building, must be installed in a readily accessible location and be protected from corrosion and other damages, including, if installed outside a building, vehicular damage. Each service regulator installed within a building must be located as near as practical to the point where the pipe enters the house. Each meter installed within a building must be located in a ventilated place and not less than 914 mm from any source of ignition or any source of heat which might damage the meter.

Service regulator vents and relief vents must terminate outdoors, and the outdoor terminal must be rain and insect resistant and be located at a place where gas from the vent can escape freely into the atmosphere and away from any opening into the building. Each regulator that might release gas in its operation must be vented to the outside atmosphere. It must also be protected from damage caused by submergence in areas where flooding may occur. The valve seat in the regulator must be made of resilient material designed to withstand abrasion of the gas, impurities in gas, cutting by the valve, and to resist permanent deformation when it is pressed against the valve port.

Each underground service line installed through the outer foundation wall of a building must: if it is metal, be protected against corrosion and if it is plastic, be protected from shearing action and backfill settlement, and both should be sealed at the foundation wall to prevent leakage into the building. Each customer household must have a line valve that must be installed upstream of the regulator or meter. Each household must have a shutoff valve in a readily accessible location that, if feasible, is outside of the building.

6.1.7 Test Requirements

As with any pipeline, including a local biomethane grid, should be properly tested in accordance with an appropriate standard to substantiate the maximum allowable operating pressure and find and eliminate each potentially hazardous leak. In conducting tests, it is important that each operator ensure that every reasonable precaution is taken to protect employees and the general public during the testing.

6.1.7.1 Test Medium

The test medium should be liquid, air, or an inert gas. If the pipeline, except for plastic pipes, is operated at or above 7.0 kPa_{gage} then it should be tested to at least 620 kPa_{gage}. For plastic pipe, the test pressure should be 150% of the maximum operating pressure. All of the joints must also be leak tested at these pressures. During construction, the pipe should be laid down in its trench with accessibility to all joints that pose a potential leak threat.

The following is an example of an appropriate test procedure:

- Pressurize the system slowly to 80% of the test pressure.
- Check for initial leaks with a bubble leakage test.
- Pressurize to the final test pressure and allow the system to stabilize for 15 min.
- Monitor the pressure for at least 2 h. There should be no noticeable pressure drop.

If the operating pressure is below 7.0 kPa gauge then the test pressure is recommended to be 70 kPa. The testing procedure is similar, which is given below:

- Pressurize to the final test pressure and allow the system to stabilize for 5 min.
- Monitor the pressure for at least 15 min.
- There should be no noticeable pressure drop.

For thermoplastic materials, the temperature during leak test should not be more than 38 °C.

6.1.8 Records

It is recommended that each operator makes and retains for the useful life of the pipeline, a record of each test performed. The record should contain at least the following information:

1. The operator's name, the name of the operator's employee responsible for making the test, and the name of any test company used.
2. Test medium used.
3. Test pressure.
4. Test duration.
5. Pressure recording charts or other records of pressure readings.
6. Elevation variations, whenever significant for the particular test.
7. Leaks and failures noted and their disposition.

6.1.9 Operations

The correct operation of a pipeline is important, especially in emergency procedures when operations go astray. With that in mind, each operator should prepare a manual of written procedures for conducting operations and maintenance activities and for emergency responses. This manual should be reviewed and updated by the operator at least once each year. This manual should be prepared before operations of the biomethane pipeline system commence. The manual should be kept at locations where operations and maintenance activities are conducted. Each operator should define the roles and responsibilities of a controller during normal, abnormal, and emergency operating conditions. There should be procedures for surveillance to determine and take appropriate action if there are failures, leakage history, corrosion, substantial changes in cathodic protection requirements, and other unusual operating and maintenance conditions.

When biomethane flows through the pipeline for the first time, it will naturally purge the pipe of the air inside. If the air and biomethane together form a hazardous mixture inside the pipe then a slug of inert gas must be released into the line before the biomethane.

6.1.9.1 Abnormal Operation

Abnormal operations are any operations where the operating design limits have been exceeded. There should be procedures identified for the following situations:

1. Responding to, investigating, and correcting the cause of:
   (i) Unintended closure of valves or shutdowns;
   (ii) Increase or decrease in pressure or flow rate outside normal operating limits;
   (iii) Loss of communications;
   (iv) Operation of any safety device; and
   (v) Any other foreseeable malfunction of a component, deviation from normal operation, or personnel error, which may result in a hazard to persons or property.
2. Checking variations from normal operation at sufficient critical locations in the system to determine continued integrity and safe operation.
3. Notifying responsible operator personnel when notice of an abnormal operation is received.

6.1.9.2 Emergency Plans

There should be written procedures to minimize the hazard resulting from a biomethane pipeline emergency. At a minimum, the procedures should provide the following:

1. Receiving, identifying, and classifying notices of events which require immediate response by the operator.
2. Establishing and maintaining adequate means of communication with appropriate fire, police, and other public officials.
3. Prompt and effective response to a notice of each type of emergency, including the following:

   (a) Gas detected inside or near a building.
   (b) Fire located near or directly involving a pipeline facility.
   (c) Explosion occurring near or directly involving a pipeline facility.
   (d) Natural disaster.

4. The availability of personnel, equipment, tools, and materials, as needed at the scene of an emergency.
5. Actions directed toward protecting people first and then property.
6. Emergency shutdown and pressure reduction in any section of the operator's pipeline system necessary to minimize hazards to life or property.
7. Notifying appropriate fire, police, and other public officials of gas pipeline emergencies and coordinating their responses during an emergency.
8. Safely restoring any service outage.
9. Train the appropriate operating personnel to assure that they are knowledgeable of the emergency procedures and verify that the training is effective.

Table 6.13 Inspection of pipeline inspection frequency

	Maximum interval between patrols	
Class location of line	At highway and railroad crossings	At all other places
1, 2	At least twice per year	
3	4.5 months, but at least four times each year	7.5 months, but at least twice each year
4	4.5 months, but at least four times each year	4.5 months, but at least four times each year

10. Each operator shall establish procedures for analyzing accidents and failures, including the selection of samples for the purpose of determining the causes of the failure and minimizing the possibility of a recurrence.

6.1.9.3 Odorization of Gas

Biomethane in the pipeline must be odorized so that at a concentration in air of one-fifth of the lower explosive limit, the gas is readily detectable by a person with a normal sense of smell. Odorization has been discussed already in Sect. 3.3.2.

6.1.9.4 Pipeline Inspection

For safety, there should be an inspection or patrol program to observe surface conditions on and beside the biomethane pipeline, checking for indications of leaks, construction activity, and other factors affecting safety and operation. Leakage surveys of the pipeline should also be conducted at the same time periods. Methods of inspection include walking with a gas detector held at most 50 mm above the ground, or other appropriate means of traversing the right-of-way. The frequency of patrols may not be longer than prescribed in Table 6.13:

A line marker should be placed and maintained as close as possible over a buried pipeline in two locations:

1. At every crossing of a public road and railroad and
2. Wherever necessary to identify the location of the transmission line or main to reduce the possibility of damage or interference.

The following should be written clearly and legibly on each line marker: The words "Warning", "Caution", or "Danger" followed by the words "Biomethane Pipeline". The letters must be at least 25 mm high. Second, the name of the operator and the telephone number where the operator can be reached should be written underneath.

6.1.9.5 Leakage Surveys

Each operator should conduct periodic leakage surveys. The exact type and scope of the leakage control program must be determined by the nature of the operations and the local conditions, but it must meet the following minimum requirements:

1. Class 3 and 4 locations: A leakage survey with leak detector equipment must be conducted, including tests of the atmosphere in gas, electric, telephone, sewer, and water system manholes, at cracks in pavement and sidewalks and at other locations providing an opportunity for finding gas leaks, at intervals not more than 15 months, but at least once per calendar year.
2. Class 1 and 2 locations: A leakage survey with leak detector equipment must be conducted as frequently as necessary but at least once every 5 calendar years at intervals not exceeding 63 months.

6.1.9.6 Abandonment or Deactivation of Facilities

The operator should have a clear plant for the abandonment or deactivation of the biomethane pipeline. Each pipeline abandoned must be disconnected from all sources and supplies of biomethane and purged of the gas. Whenever service to a customer is discontinued, one of the following must be complied with:

1. The valve that is closed to prevent the flow of gas to the customer must be provided with a locking device or other means designed to prevent the opening of the valve by persons other than those authorized by the operator.
2. A mechanical device or fitting that will prevent the flow of gas must be installed in the service line or in the meter assembly.
3. The customer's piping must be physically disconnected from the gas supply and the open pipe ends sealed.

6.2 Case Study

6.2.1 Local Biomethane Grid in Chiang Mai

There have been three biomethane local grid networks built in Thailand for consumption of biomethane gas in domestic cooking and heating applications. Details of the biomethane plant are provided in Sect. 3.5. Figure 6.3 shows the layout of the pipeline network in Chiang Mai with pipeline lengths and loads shown in Table 6.14. It was constructed using the technical standard described throughout this chapter. Figure 6.2 shows the pipeline under construction. The number of villagers participating in the project, totaled 120 households, with an average daily gas consumption of 0.6 kg/day. The assumptions that went into this design were as follows:

Table 6.14 Layout and diameter of biogrid pipelines for different outlets

Section	Number of households/section	Section length (m)	Distance from plant (m)	Run length (m)	Load delivered/section (m^3/hr)
BG-001	0	250	250	1935	45.0
BG-002	5	160	410	410	6.0
BG-003	5	300	710	710	1.9
BG-004	6	222	632	632	2.25
BG-005	8	250	500	1935	39.0
BG-006	3	180	680	680	7.5
BG-007	5	100	780	780	1.9
BG-008	9	200	880	880	3.4
BG-009	3	140	820	820	1.1
BG-010	2	100	880	1935	28.5
BG-011	3	100	980	980	1.1
BG-012	3	150	1030	1935	26.6
BG-013	3	180	1210	1210	1.1
BG-014	0	100	1130	1130	4.1
BG-015	1	120	1250	1250	3.75
BG-016	4	100	1350	1350	1.5
BG-017	5	90	1340	1340	1.9
BG-018	1	140	1270	1270	0.4
BG-019	5	220	1250	1935	20.25
BG-020	7	330	1580	1580	2.6
BG-021	8	185	1435	1935	15.75
BG-022	2	220	1655	1655	0.75
BG-023	0	30	1465	1935	12.0
BG-024	7	290	1755	1755	2.6
BG-025	2	120	1585	1935	9.4
BG-026	1	60	1645	1645	4.1
BG-027	2	60	1705	1705	0.75
BG-028	8	140	1785	1785	3.0
BG-029	3	140	1725	1725	1.1
BG-030	9	350	1935	1935	3.4
Total	120	5027			

- Each house cooks for 2 h per day and uses biomethane at a rate of 0.375 Nm^3/hr.
- The upstream or inlet pressure from the plant is 500 kPa_{abs}.
- To ensure constant gas delivery, the pressure at the farthest point of use should be kept within 10% of the supply pressure.
- The specific gravity of the gas is 0.667.

Fig. 6.2 Construction of biogas grid pipeline

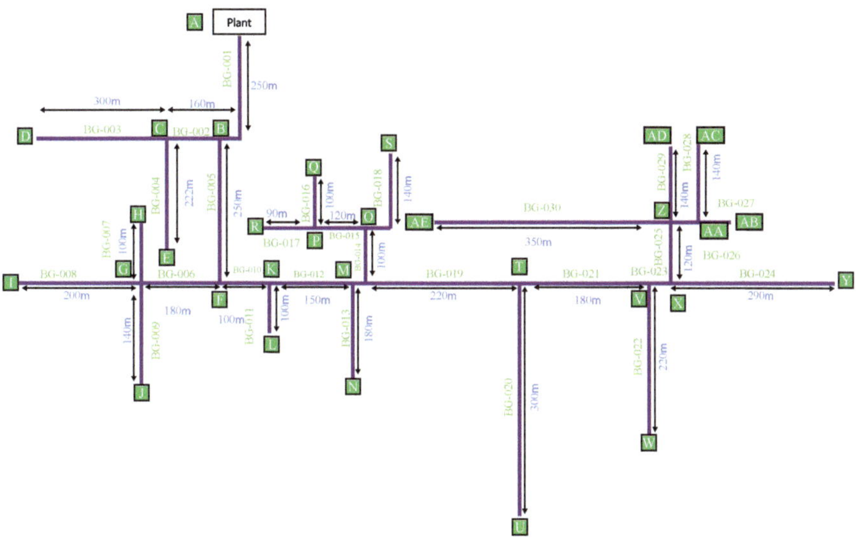

Fig. 6.3 Layout of pipeline for biogrid system

In order to calculate the pipe diameter necessary for transporting the gas, the Mueller equation (Eq. 6.4) is used. Having a pipe diameter too large adds unnecessary cost while having it too small results in too great a pressure drop.

$$Q = \frac{2826 D_{in}^{2.725}}{S_g^{0.425}} \left[\frac{P_1^2 - P_2^2}{L} \right]^{0.575} \tag{6.4}$$

Table 6.15 PE 100 HDPE pipe dimensions

OD (PN10) (mm)	ID (PN10) (mm)
25	20.1
32	27.1
40	34.8
50	43.7
63	55.0
75	65.6

where Q = Gas flow rate in CFH
S_g = Gas specific gravity
P_1 = Inlet pressure lb/in^2 absolute
P_2 = Outlet pressure lb/in^2 absolute
L = Length, ft
D_{in} = Pipe inside diameter, in

Table 6.15 shows the commercially available dimensions for polyethylene pipe. It has a maximum operating pressure of 1 MPa at 20 °C and a Standard Dimensional Ratio (SDR), outside diameter divided by the wall thickness of 17. They conform to standards ISO 4427 and ISO 4065. Table 6.16 shows how the inner diameter of each pipeline section was calculated using the Mueller equation (Eq. 6.4). The calculated diameter was then rounded up to the nearest commercially available diameter, Table 6.15. It can be seen that the maximum needed polyethylene pipe diameter was 50 mm with the smallest being 25 mm. Components used during the construction and installation can be seen in Fig. 6.4.

The total pressure drop through this system was kept to 2.8%, well below the design criteria of 10% helping to ensure a constant and smooth flow of gas to all households.

Safety is a major concern and at the beginning of each section a shutoff valve is installed with a drain valve installed at the exit of each section, Fig. 6.5. Even though the biomethane contains only a small amount of moisture, over time the moisture can accumulate and condense out. Excess flow valves are located with every shutoff valve. Excess flow valves are designed to automatically shut off the gas flow in the event of a catastrophic leak. These valves do not prevent against slow leaks. A service team consisting of 15 local volunteers is responsible for the day-to-day operation of the pipeline. They remove excess pipeline moisture on a monthly basis using the drain valves. They are responsible for billing and gas fee collection. Every household has its own gas meter and receives bills based on monthly gas consumption. The rate is set by the locals.

Table 6.16 Layout and diameter of biogrid pipelines for different outlets

Section	Load delivered/section (m^3/hr)	P (inlet) kPa_{abs}	P (outlet) kPa_{abs}	Inner diameter from Eq. 6.4 (mm)	HDPE Pipe size OD from Table 6.15 (mm)
BG-001	45.0	500	495.2	41	50
BG-002	6.0	495.2	491.6	15	25
BG-003	1.9	491.6	490.7	11	25
BG-004	2.25	491.6	490.6	12	25
BG-005	39.0	495.2	491.5	41	50
BG-006	7.5	491.5	490.1	19	32
BG-007	1.9	490.1	489.8	12	25
BG-008	3.4	490.1	488.4	15	25
BG-009	1.1	490.1	489.9	10	25
BG-010	28.5	491.5	490.6	38	50
BG-011	1.1	490.6	490.5	11	25
BG-012	26.6	490.6	489.5	39	50
BG-013	1.1	489.5	489.3	12	25
BG-014	4.1	489.5	489.2	18	32
BG-015	3.75	489.2	488.9	19	32
BG-016	1.5	488.9	488.7	14	25
BG-017	1.9	488.9	488.7	16	25
BG-018	0.4	489.2	489.2	8	25
BG-019	20.25	489.2	488.2	38	50
BG-020	2.6	488.2	487.7	18	32
BG-021	15.75	488.2	487.6	37	50
BG-022	0.75	487.6	487.5	12	25
BG-023	12.0	487.6	487.5	35	50
BG-024	2.6	487.5	487.2	20	32
BG-025	9.4	487.5	487.4	33	50
BG-026	4.1	487.4	487.2	25	32
BG-027	0.75	487.2	487.2	14	25
BG-028	3.0	487.2	487.0	23	32
BG-029	1.1	487.4	487.2	16	25
BG-030	3.4	487.4	486.7	24	32
Largest pressure drop			13.8 kPa (2.8%)		

6.2.2 *Economics of the Biomethane Grid*

Pipeline construction costs depend on the length of pipeline which is dependent on the number of households serviced. The cost of building this biogas plant was roughly

(a) Pressure reducing system

(b) Excess flow valve

(c) Odorization system, see Sect. 3.3.2

(d) Gas meter cabinets

Fig. 6.4 Photos of different components used throughout the biomethane grid network

$130,000, the upgrading plant (water scrubbing) was $212,000 with the pipeline costing $268,000 for a total of $610,000. This system has a capacity to serve 500 households but serves only 120 in this community. For 500 households, the annual operating costs, including labor and electricity, run to $30,000 approximately. If the gas is sold at a price of $0.54/kg and assuming an interest rate of 6.5% and the end user is responsible for their own biomethane stoves (see Sect. 5.3.1), the internal rate of return for this project works out at 7.4%.

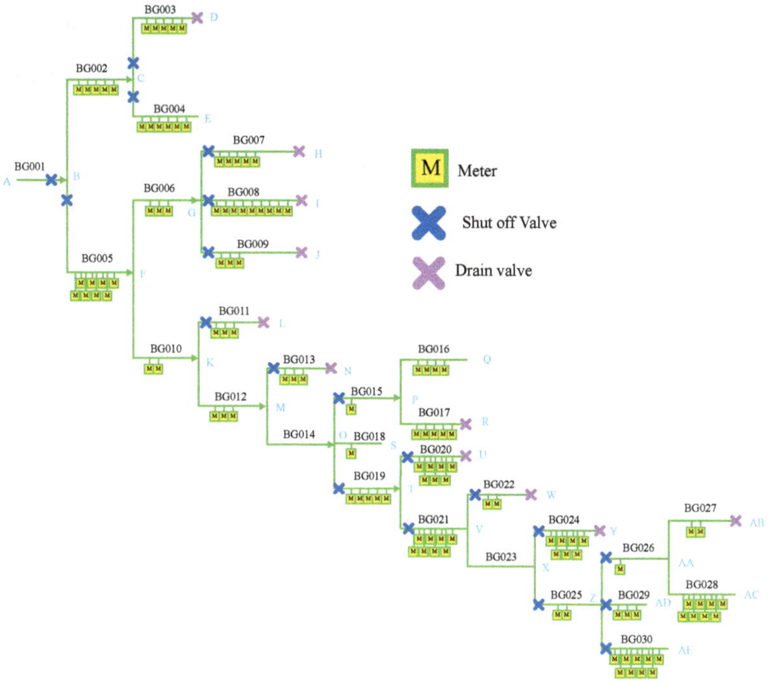

Fig. 6.5 Meters, shutoff valves, and drain valves throughout the local gas pipeline

References

1. American Society of Mechanical Engineers (2004) Gas transmission and distribution piping systems. https://law.resource.org/pub/us/cfr/ibr/002/asme.b31.8.2003.pdf (Standard ASME B31.8)
2. Moran J, Koonphapdeelert S, Bunkham A, Rojanaratanangkule W, Aggarangsi P (2016) Development of a national standard for a biogas grid in Thailand. J Environ Sci 5:1–7
3. International Standards Organization (2013) Natural gas - quality designation. http://www.iso.org/iso/catalogue_detail.htm?csnumber=53058 (ISO 13686:2013)
4. Rasi S (2009) Biogas composition and upgrading to biomethane. Master's thesis, University of Jyvaskyla, Jyvaskyla, Finland
5. European Committee for Standardization (2016) Natural gas and biomethane for use in transport and biomethane for injection in the natural gas network - part 1: specifications for biomethane for injection in the natural gas network
6. Svensson M (2014) Biomethane standards. In: European biomethane workshop, Brussels
7. Roggenkamp MM, Tempelman DG (2012) Looking back, looking ahead: gas sector developments in The Netherlands and the eu: from manufactured gas via natural gas to biogas. J Energy Nat Res Law 30(4):523–537
8. European Biogas Association (2019) Biomethane: responsibilities for injection into natural gas grid. Technical report, Marcogaz
9. City Gas PTE (2013) Handbook on gas supply. http://www.citygas.com.sg/pdf/City_Gas_Handbook_on_Gas_Supply_Nov13.pdf
10. Suwansri S, Moran J, Aggarangsi P, Tippayawong N, Bunkham A, Rerrkriangkrai P (2014) A biomethane solution for domestic cooking in Thailand. Energy Sustain Dev 23:68–77

11. EASEE-Gas (2005) Harmonisation of natural gas quality. Technical report, European association for the streamlining of energy exchange, Paris
12. Euromot (2011) Gas quality aspects for reciprociating gas engines. Technical report, the European association of internal combustion engine manufacturers, Frankfurt
13. ERDI (2013) A prototype bio-methane gas compressor for automotive applications (in Thai). Technical report, energy research and development institute, Chiang Mai
14. Slim BK, Darmeveil HD, Gersen S, Levinsky HB (2011) The combustion behavior of forced draught industrial burners when fired with the easee-gas range of wobbe index. J Nat Gas Sci Eng 3:642–645

Chapter 7
Biomethane—Future Trends

7.1 Biogas Production

Biomethane, the renewable alternative to natural gas, is principally produced from two complementary technologies: anaerobic digestion and gasification. Speirs et al. [1] conducted a review of network decarbonization options. The focus was on technologies used to generate biomethane and hydrogen. Ignoring the options involving carbon capture and storage which are unproven at large scale, the lowest carbon option was alkaline water electrolyzer fed by wind with the next lowest being bio gasification. This study agrees with the data shown in Fig. 4.1 which supports the thesis that for certain substrates the GHG potential is negative for biomethane.

Anaerobic digestion is becoming a key renewable energy source in Europe which in 2014 counted 17,240 plants and over 70,000 people working in the sector. The global production trend can be seen from Fig. 1.4. In 2013, Europe produced the natural gas equivalent of 15.6 billion Nm^3 [2], while that year's total EU gas consumption was 472 billion m^3 [3]. Note that this comparison is only theoretical, since just a small fraction of 15.6 billion m^3 was upgraded into biomethane, where over 90% of total energy was used in the form of raw biogas (not upgraded to natural gas quality) to produce electricity and heat in cogeneration units. Although this is only a small share of the total biogas production, several factors are encouraging biogas producers to make the switch to biomethane upgrading, particularly increasing electricity prices, scarcity of fossil fuels, and commitments to reduce GHG emissions.

Gasification technology has been developing rapidly in several reputed research centers and innovative companies across Europe. Although gasification has not yet been deployed at a large industrial scale, its high efficiency rates compared to incineration make it a very interesting environmentally friendly option. This technology complements anaerobic digestion very well, as it can turn feedstock that is impossible to digest such as woody biomass and polluted organic waste into biomethane. This cutting edge offers the best solution for the management of various organic materials and amplifies the potential of renewable energy.

S. Koonaphapdeelert et al., *Biomethane*, Green Energy and Technology,
https://doi.org/10.1007/978-981-13-8307-6_7

Table 7.1 Substrate potential estimations in billion m^3/year for biomethane, (Reprinted with permission from [4])

Origin	Production potential 2019 (m^3/year)
Animal manure	37.9×10^9
Livestock effluents	3.6×10^9
Fruit by-products	1.1×10^9
Vegetable by-products	15.1×10^7
Energy crops	61.5×10^9

Table 7.1 shows the production potential in 2019 if all organic wastes in Europe were processed into biomethane, from [4]. The total potential is 104 billion m^3/year or about 22% of Europe's gas consumption. The total maximum technical potential for renewable gas that can be achieved in Europe through the combination of anaerobic digestion and gasification is estimated to be 151–246 billion m^3, where the lower end represents a pathway with low energy crop use and the upper one a more intensive deployment of dedicated crops. Note that this technical potential cannot be realized in practice but it provides an indication of the magnitude of possibilities in Europe. Animal (bovine and pig) manure and energy crops are the substrates that have the most energy potential. The methane yield for livestock effluents is not high but the vast quantity of manure produced allows high annual methane production. Energy crops (barley, maize, wheat) have a high energy density, quick kinetics, and large production potential. Not all of these potential sources are economical. The Green Gas Grids project [5] concluded that the maximum Europe can economically produce is 48–50 billion m^3 of natural gas equivalent by 2030 from gasification and anaerobic digestion. This prediction is in line with the overall commitments of each EU country's National Renewable Energy Action Plan, which amounts to 28 billion m^3 of biogas/biomethane by 2020 [6].

7.1.1 Acid-Phase Digesters

In the future, technological developments should increase the rate of biomethane production by increasing the organic loading rate. This allows for more waste to be converted to biogas for a given digester size. One promising technology uses an acid-phase digester. Inside a typical anaerobic digester, there are four stages:

1. Hydrolysis: Degradation of complex high-molecular-weight organic macromolecules into low-molecular-weight monomers.
2. Acidogenesis: Conversion of soluble monomers to various metabolic products.
3. Acetogenesis: Conversion of products from acidogenesis to acetic acid and hydrogen.
4. Methanogenesis: Methane production.

All involve breaking down biodegradable materials using a variety of microorganisms. During first stage, hydrolysis, complex organic compounds are converted to Volatile Fatty Acids (VFAs). This stage is characterized by low Hydraulic Retention Time (HRT), acidic pH, and high volatile acid concentrations. The hydraulic retention time is the average length of time that the organic compounds remain in the digester.

$$HRT\ (day) = \frac{Digester\ Capacity\ (L)}{Daily\ Substrate\ Added\ (L/d)} \tag{7.1}$$

A well-operated digester requires the rates from all stages to be balanced. For instance, if the hydrolysis stage runs too fast, the VFA concentration significantly increases, and the pH quickly drops below 7. The excessive buildup of VFAs inhibits the fourth stage, methanogenesis. However, the hydrolysis stage can progress at a faster rate than the remaining three stages [7]. Therefore, as the organic loading increases, the risk of inhibition due to excessive VFA production also increases. This is where introducing an acid-phase stage prior to the main digester comes in. It is in effect a separate digester in which only hydrolysis occurs. The purpose of the acid-phase digester is to provide a separate environment for the hydrolysis microorganisms. The acid-phase digester could be operated at different parameters to optimize hydrolysis. There is evidence to suggest that a higher temperature (55–70 °C) is more suited to hydrolysis than methanogenesis. Another factor warranting separation of digesters is the pH level. Each reaction rate has a different optimal pH level. For example, methanogenesis is most efficient at a pH between 6.5 and 8.2 [8]. The rate of methanogenesis greatly reduces at pH levels below 6.6. The optimum pH of acidogenesis is between pH 5.5 and 6.5 [9]. Separation allows optimization of each process. Adding a large volume of new material daily may result in changes to the digester's environment. A large Organic Loading Rate (OLR) can lead to higher hydrolysis/acidogenesis bacterial activity compared to methanogenesis bacterial activity.

Nghiem et al. [10] found that acid-phase digestion pretreatment resulted in an increase in biogas production and volatile solids (VS) removal at a sewage treatment plant in New South Wales, Australia. Without the acid-phase digesters, the specific methane yield of the waste activated sludge was $190\,L/kg_{VS}$, whereas a specific methane yield of $231\,L/kg_{VS}$ was observed from sludge from the acid-phase digester.

7.1.2 Anaerobic Digestion of Microalgae

Biogas production using algal resources has been widely studied as a future green and alternative renewable technology. These are called second-generation biofuels. Algae are diverse category of photosynthetic entities from different phylogenetic groups and are classified as microalgae and macroalgae (also known as seaweeds; brown, green, and red algae). Algae have been recommended as second-generation renewable feedstock for sustainable biogas production. They exhibit various advantages in higher photosynthesis efficiencies (around 3–8%) which is substantially higher than that of terrestrial plants (typically 0.5%). Since they are marine based they have less or no land requirement as compared with land-based plants [11]. Microalgae has less potential to interfere with food production and have been found to adapt in stressful conditions. The gas production efficiency has been shown to be species dependent. One factor is the differences in structure of the microalgae cell walls. Many microalgae species have difficulty hydrolyzing. Thus, pretreatment is necessary to disrupt or solubilize the cell walls. Generally, pretreatment methods are species specific and their success depends on the nature of the cell wall. Saratale et al. [12] provide an extended overview of recent advances in biomethane production via direct anaerobic digestion of microalgae and future challenges for its scaled-up applications. There is a lot of existing literature on lab-scale algal biomass. However, very few large-scale applications have been reported. Wu et al. [13] estimated a biomethane production cost of $\$0.66/kg_{Biomethane}$ from algae in a covered anaerobic lagoon which is expensive compared with more traditional production methods, see Table 7.3. In future, large-scale, cost-effective production facilities based on microalgae will be used for biomethane production. Xiaoqiang et al. [14] increased the annual biomethane output by 9.4% in a Swedish biogas plant by integrating algae cultivation with the plant. The algae were cultivated in a greenhouse photobioreactor. The discharges of CO_2 and digestate from the gas production were fed back as nutrient input to the photobioreactor. The authors claimed the process could be scaled and used in future designs.

7.1.3 *Lignocellulosic Biomass*

Lignocellulosic biomass is another second-generation biofuel. Lignocellulosic biomass inherited their name from their main components, lignin (4–35% of total solids), hemicellulose (10–37% of total solids), and cellulose (17–61% of total solids). It is a biomass that can be divided into four categories: hardwood (eucalyptus, oak), softwood (spruce, pine), agricultural wastes (cotton stalk, nutshells, rice husk, sugarcane, corn stover), and grasses (bamboo, hemp, rye, reeds). Lignin, a main component of cell walls, provides structural rigidity and resistance against oxidative stress and microbial attack. The major issue is their resistance to digestion. It cannot be digested because the lignin, cellulose, and hemicellulose have low biodegradation and poor digestion performance. Extra measures are needed to start the digestive process such as pretreatment. The current process requires pretreatment with severe conditions to disrupt the plant cell wall structures and remove hemicellulose and lignin components. So, there is a need to develop more suitable pretreatments prior to digestion. Various pretreatments, such as physical, chemical, and biological methods, have been employed to improve biogas production from lignocellulosic biomass [11]. Future research will target eco-friendly, economical, and time-effective solutions.

One approach is to use fungi in the pretreatment of lignocellulosic biomass, particularly those that attack lignin. Some studies show that after fungal pretreatment, a 5 to 15% increase in the methane yield can be obtained [15, 16]. Another approach is to use a microbial consortium. This consortium contains yeast and cellulolytic bacteria. It can increase digestibility. Some studies have reported methane yield improvements of 25–96.63% by using microbial consortia [17]. Another method to remove the hemicellulose is to use an alkali such as sodium hydroxide in pretreatment. This can break the links between lignin monomers. As a result, the specific surface area is increased [18]. Alkalis help to prevent decrease in pH during the acidogenesis process, increasing the efficiency of methanogenesis. Adding iron sulfate ($FeSO_4$) has been shown to aid biogas production in cow manure and poultry litter anaerobic digesters [19]. Iron (Fe) reacts with H_2S to form FeS; therefore, the addition of iron can be used to in situ reduce the toxicity of H_2S in biogas. Other emerging technologies include ionizing and nonionizing radiation, pulsed electrical fields, ultrasound, and high hydrostatic pressure which may provide the breakthrough needed in the valorization of lignocellulosic biomass.

7.2 Biogas Upgrading

In addition to the commercially available techniques mentioned in Chap. 3, there are several new technologies under development. Each helps to purify biogas by removing carbon dioxide and other impurities.

7.2.1 Cryogenic Upgrading

Cryogenic upgrading makes use of the different boiling/sublimation points of the different gases. It can produce biomethane that is 100% pure. In applications where liquid biomethane (LBM) is used, see Sect. 7.3.2, this process can both purify and liquefy the methane. The raw biogas is cooled down to the temperatures where the carbon dioxide in the biogas condenses and can be separated as a liquid, while methane remains in the gas phase. At atmospheric pressure, CH_4 condenses at $-161.6\,°C$ and CO_2 freezes at $-78.5\,°C$. This makes it easy to separate. However, when mixed with methane, a lower temperature is required to make the carbon dioxide condense [20]. This technology can also extract moisture and siloxanes simultaneously. It is best performed at elevated pressure to ensure that CO_2 condensates into a liquid and not a solid form. If it is a solid (dry ice) that would clog the piping system. If the biogas contains a lot of nitrogen then condensing the CH_4 will also remove the nitrogen.

There are several steps to this process. The first phase occurs between 1.7 and 2.6 MPa with the mixture cooled to $-25\,°C$, the moisture, hydrogen sulfide, and sulfur dioxide are removed at this step. Siloxanes, if present, are also removed here. The remaining gas then flows through a filter, removing contaminants. The second stage, which is cooled between −50 and −59 °C, causes condensation of carbon dioxide. If desired the temperature can be further cooled, with the carbon dioxide becoming a solid and the end product is methane in a liquid form.

Manufacturer's of cryogenic upgrading units include Acrion Technologies, Air Liquide, Cryostar, FirmGreen, Gas treatment Services, and Gasrec.

7.2.2 In Situ Methane Enrichment

The in situ methane enrichment technique is performed by pumping the digester sludge which is rich in soluble CO_2, through a CO_2 desorption column. The desorption of CO_2 is achieved by aeration of the sludge. After it is then pumped back to the digester. CO_2 has a higher solubility in water compared to CH_4. It is 20 times more soluble than CH_4, at a pH of 7.0 and a temperature of 35 °C [21]. The returning sludge will not be saturated in CO_2, and will be able to absorb more CO_2 produced in the digester. It will also absorb some CH_4 but this will be a much smaller fraction of the biogas.

It is possible to produce methane at concentrations as high as 95%, with a loss of just 2%. The process costs only one-third of the conventional upgrading process of biogas and is relatively simple to implement. However, its limitation is that it cannot produce methane that exceeds 95% in concentration and the sludge has to have a low enough viscosity to pump.

7.2.3 Upgrading Biogas with Enzymes

This process uses natural enzymes, carbonic anhydrase, which are present in our bloodstream. They catalyze carbon dioxide and water into bicarbonate and protons through a reversible reaction, Eq. 7.2. Through metabolism these enzymes help to transport carbon dioxide out of tissues.

$$H_2O + CO_2 \longleftrightarrow H^+ + HCO_3^- \tag{7.2}$$

The carbon dioxide, which is in the form of carbonate (HCO_3^-) is taken to the air sacs in the lungs. Research is ongoing to improve biogas quality using carbonic anhydrase. A Canadian company CO_2 Solution Inc. has developed a patented bioreactor that uses enzymes.

7.2.4 Future Developments in PSA Technologies

In PSA technology, the focus is on making the process more energy efficient, suitable for small-scale applications, more efficient adsorbents, and segregating H_2S and CO_2 in a single column [22]. One of the biggest issues with PSA is the CO_2 stream which contains a significant amount of methane [23]. This must be cleaned before emitting to the environment which increases the cost. Adsorbent development is one of the most important areas of research for improving PSA technology. Novel adsorbents like Metal-Organic Framework (MOFs) and Zeolitic Imidazolate Frameworks (ZIFs) may offer working capacity and selectivity advantages.

7.2.4.1 Small-Scale Upgrading Units

Companies that focus on small-scale biogas upgrading include BioGTS, Biosling, Biofrigas, BMF Haase Energietechnik, Metener, NeoZeo, and Sysadvance.

7.3 Biomethane Applications

7.3.1 Low-Pressure Biomethane Storage

In addition to improvements in biomethane production, there will be improvements in how biomethane is used in the future. One potential future technology is low-pressure biomethane storage which allows for a larger mass of biomethane to be stored than predicted from the ideal gas law. The natural gas in vehicles (NGV or CNG) must store the gas at high pressures, up to 20 MPa in order to travel a reasonable distance without refilling. For example, taking a regular 28 liter tank at 20 MPa and 300 K, from the ideal gas law, it contains

$$\frac{PV}{R_{CH_4}T} = \frac{20 \times 10^6(28 \times 10^{-3})}{\frac{8314}{16}(300)} = 3.6\,\text{kg} \tag{7.3}$$

about 3.6 kg of methane, see Eq. 7.3. This corresponds to around 193 MJ of energy. A regular 12-gallon automotive gasoline tank contains 1500 MJ of energy. Filling a CNG tank also requires an expensive multistage compression facility. In addition to safety concerns, the necessary tank wall thickness is expensive and heavy. There would be significant advantages to storing an equal amount of gas, in the same volume but at lower pressures. Adsorbed Natural Gas (ANG) is a promising technology which offers a path to this solution [24]. In Sect. 3.2.2, this adsorption principle was introduced in the context of upgrading biogas using pressure swing adsorption.

ANG is based on adsorption technology which is the adhesion of atoms or molecules from a liquid or gas, the adsorbate, to a surface, the adsorbent. This process creates a film of the adsorbate on the adsorbent. Atomic bonds between adjacent layers of the adsorbate and adsorbent hold the adsorbate in place. Important adsorbent parameters are the specific surface area (m^2/g) and the average pore size. ANG is natural gas which is stored at relatively low pressures in a tank filled with an adsorbent. Known adsorbents include charcoal, silica, and treated alumina. Chiew

Table 7.2 Properties of the absorbents used (Reprinted with permission from [24])

Absorbent	Specific surface area (m^2/g)	Total pore volume (ml/g)	Average radius of porosity (Å)	Percentage micropore (less than 20 Å) (%)	Bulk density (g/ml)
Activated carbon 1	807.8	0.344	8.53	>70	0.485
Activated carbon 2	1,574.7	0.876	11.13	>70	0.309
Activated carbon 3	2,365.9	1.677	14.18	>70	0.312
Activated Alumina	316.6	0.462	29.18	>70	0.742
Molecular Sieves 13x	608.8	0.342	11.24	<50	0.633

et al. [25] demonstrated that ANG can contain two to three times more natural gas at one-sixth of the normal CNG pressure.

Sirichai et al. [24] carried out biomethane adsorption experiments on activated carbon 1, activated carbon 2, activated carbon 3, alumina, and zeolite 13x. Their relevant properties are shown in Table 7.2. Activated carbon 1 and 2 were synthesized from organic coconut shells. In this process, all organic materials were entirely removed from the elemental carbon structure. Carbon layers react to produce carbon dioxide and hydrogen gas, leaving behind open pores. The activated Carbon 3 came from pinewood and can be purchased commercially.

Testing was carried out where the quantity of biomethane in the tank was measured as a function of the tank pressure. Figure 7.1 shows that per kg of adsorbent activated carbon 3 adsorbed the most biomethane at a given pressure, followed closely by activated carbon 2. The adsorption capacity was not linearly correlated with the specific surface area. Activated carbon 3 had nearly 50% excess surface area over activated carbon 2 but only adsorbed 3% more biomethane at 5 MPa. The pore volume of activated carbon 3 was 1.677 ml/g, the highest among all adsorbents, indicating a large volume for gas storage. Alumina had a low surface area of only 316.6 m^2/g and had a large average pore size, 29.18 Å. This resulted in the lowest adsorption capacity of all five types of adsorbents.

Figure 7.2 shows the mass of biomethane contained in a liter of the gas tank. Activated carbon 3 can store the highest quantity of biomethane gas. At 5 MPa, the quantity of biomethane was 103 g/L, followed by activated carbon 2 and activated carbon 1 with 91.0 g/L and 79.5 g/L, respectively. Without any absorbent, the biomethane stored was only 50 g/L. It was found that the adsorption capacity decreased slightly with increasing temperature.

In order to see if the adsorption capability degraded after filling and emptying the tank many times, activated carbon 2 was used to test if the performance degraded with number of cycles. The tests were carried out under constant temperature conditions of

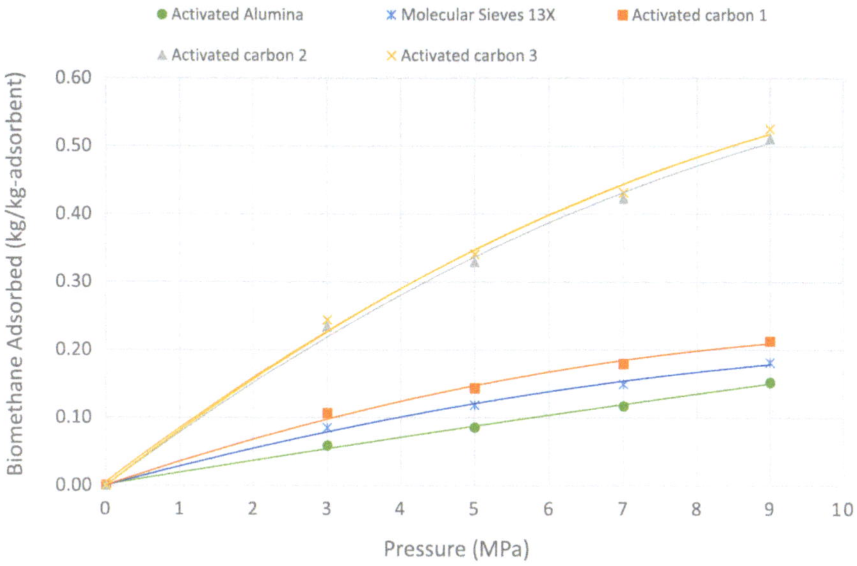

Fig. 7.1 Biomethane adsorption relative mass: (kg biomethane gas/kg adsorbent) versus adsorbent pressure at 15 °C (Reprinted with permission from [24])

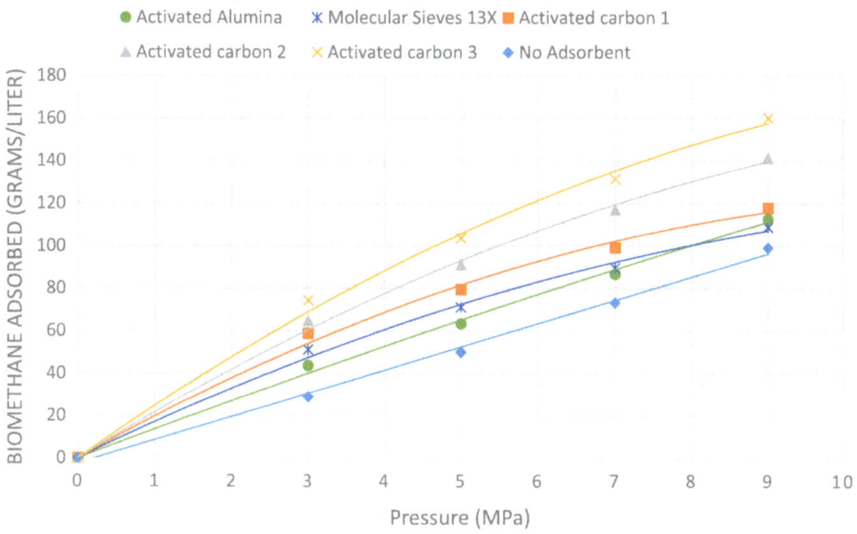

Fig. 7.2 Biomethane adsorption: (grams of biomethane per tank liter) versus pressure at 15 °C (Reprinted with permission from [24])

15 °C and the results are displayed in Fig. 7.3. In this figure, the percentage of methane in the biomethane is plotted against the number of cycles. Prior to adsorption the methane quantity was 85.0% by volume. After one cycle, the methane composition

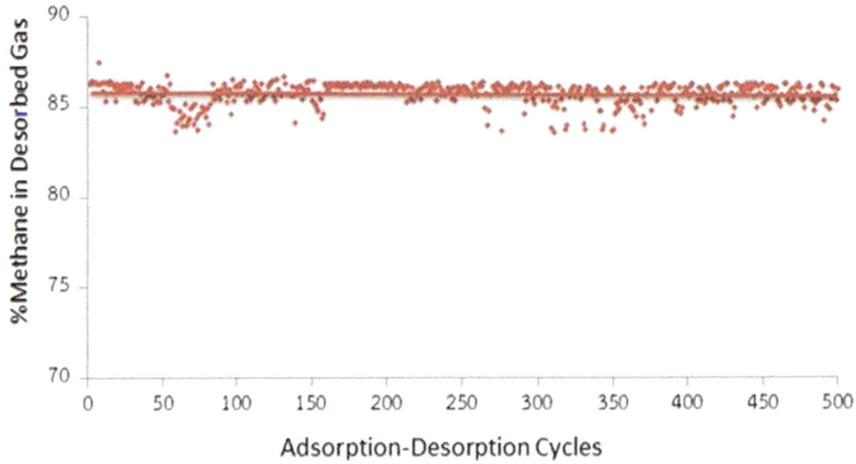

Fig. 7.3 Methane composition (CH_4) after desorption (% by Vol) versus number of cycles (Reprinted with permission from [24])

of the exit gas was 86.2% and after 500 cycles the methane percentage was 85.6%. This is a very narrow range and shows almost no degradation over 500 cycles.

The US Department of Energy has set a standard for the storage capacity of methane on adsorbent materials at 180 times the volume of methane for a given adsorbent volume. The conditions for this target are 3.5 MPa and room temperature (298 K). This means that the volume of biomethane at standard temperature and pressure (298 K, 1 atm) per volume of tank (V_{bio}/V_{tank}) should be 180 [26]. The focus of the US department of Energy was transportation applications. In the study presented above, at 3.5 MPa, the best performance was 85.5 V/V from activated carbon 3 indicating that a lot more research is needed to meet the US standard but progress is heading in the right direction. Although activated carbon 3 may not yet be suitable adsorbent for vehicular gas storage, such capacity could be adequate for household cooking gas storage which requires less energy density.

7.3.2 Biomethane Liquefaction

Another potential future use for biomethane is to liquefy it. In a liquefied state, the energy density is 600 times that of biomethane at atmospheric pressure and 3 times that of biomethane at 20 MPa. This would be useful in transportation applications and in situations were injection into a natural gas grid is not possible. The main issue with liquefying biomethane is the cost. There are three primary uses for liquefied biomethane (LBM). It can be used as a fuel replacement for NGV which means that it must be heated to a gaseous state before selling. It could also be used in specialized vehicles that use LNG or it could be used in industry to replace fuel oil, coal, or LPG.

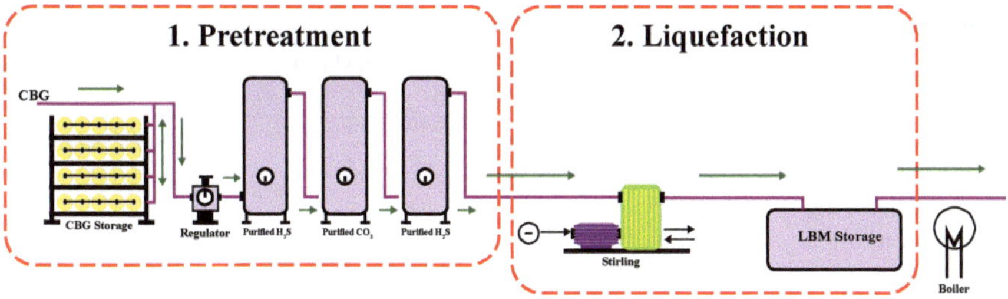

Fig. 7.4 Schematic for liquid biomethane production

For details on the cryogenic process please see Sect. 7.2.1. Presently, there are biomethane liquefaction plants operating in Sweden, Norway, UK, the Netherlands, USA, and the Philippines. In Sweden, the Lidköping Biogas Plant has a 1.2 m^3 liquefied biomethane per hour (LBM/hr) production capacity. A pilot LBM plant in the Netherlands opened in 2009. A plant built in British Columbia, Canada in 2000 produces liquid methane at a concentration of 96%. Another plant in Bowerman, California produces liquid methane at a rate of 19 m^3 per day. The Cryo Pur plant in Northern Ireland continuously produces about 3 tones per day of liquefied biomethane. A small-scale theoretical LBM production process is shown in Fig. 7.4. It is presently under design in Thailand and will be capable of producing LBM at a rate of 10 kg/hr at a temperature of $-120\,°C$ and a pressure of 20 MPa. It is not yet in operation but will be an additional add-on to the plant discussed in Sect. 1.6. The methane content in this process is greater than 90%. A Swedish study concluded that for transport distances over 200 km, liquefying the biomethane was economical while it should be left in the compressed biomethane state for distances less than 200 km.

An economic cost comparison based on costs in Thailand is shown in Table 7.3. They show that LBM production costs are higher than CBG and slightly higher than LNG import costs (at 0.33 USD/kg) but cheaper than importing LNG via long-term contracts (0.44–0.0.53 USD/kg). The cost of LBM begins to have a cost advantage over CBG if it must be transported over distances beyond 200 km. Both LBM and CBG costs are cheaper than the liquid petroleum (LPG) price of $0.7/kg, see Fig. 5.18.

Pellegrini et al. [27] modeled three natural gas low-temperature purification technologies and conducted an energy consumption analysis for LBM production. In one process, called a dual pressure process, there are two distillation units: the first one is operated at high pressure (5 MPa), while the second one at low pressure (4 MPa). The low pressure is below the methane critical pressure which is around 4.6 MPa. It was found that the dual pressure low-temperature distillation scheme resulted in the lowest equivalent methane requirement.

Research is also ongoing to convert methane into a liquid such as methanol. Current conversion processes are usually accomplished by costly and energy-intensive steam reforming at elevated temperature and high pressure. There is ongoing research to find efficient, cost-effective gas-to-liquid conversion processes using Metal-

Table 7.3 Cost estimation of CBG v LBM in Thailand

Process	CBG ($/kg)	CBG ($/GJ) (from Sect. 3.3)	LBM ($/kg)	LBM ($/GJ)
Biogas cost	0.12	$\$6.7/GJ_{biogas}$	0.12	$\$6.7/GJ_{biogas}$
Upgrading + compression or liquefaction cost	0.20	$\$4.44/GJ_{CBM}$	0.35	$7.81/GJ
Transportation cost for 100 km	0.09	2.00/GJ	0.02	0.44/GJ
Transportation cost for 200 km	0.19	$4.24/GJ	0.03	$0.67/GJ
Total cost for 100 km	0.42	$13.18/GJ	0.48	$15.0/GJ
Total cost for 200 km	0.50	$15.38/GJ	0.50	$15.22/GJ

Organic Frameworks (MOFs) as electrocatalysts. The goal is to use MOF to permit the methane-to-methanol conversion process to proceed at ambient temperature and pressure [28].

7.3.3 *Power to Gas*

In future, a Power-to-Gas process (PtG) chain could play a significant role in future energy systems, especially in countries with large renewable sources such as solar and wind. The excess energy produced at periods of low electric demand could be used to power electrolysis units and the hydrogen produced is converted into methane. Thus, renewable electric energy can be transformed into storable methane. The link to biogas comes from the methanation process which converts hydrogen into methane. It requires an external CO or CO_2 source. Any biomethane upgrading process produces CO_2 as a by-product. The upgrading plant also stores and must have an end use for the biomethane. The following are the basic reactions in the methanation process. The CO (Eq. 7.5) and CO_2 (Eq. 7.4) reactions are highly exothermic reactions with the consequence that the high temperatures limits their conversion rate. Low temperatures and high pressures give high CO_2 conversion rates.

$$CO_2(g) + 4H_2(g) \rightleftharpoons CH_4(g) + 2H_2O(g) \quad \Delta H_r^0 = -165.1\,kJ/kmol \tag{7.4}$$

$$CO(g) + 3H_2(g) \rightleftharpoons CH_4(g) + H_2O(g) \quad \Delta H_r^0 = -206.3\,kJ/kmol \tag{7.5}$$

$$CO_2(g) + H_2(g) \rightleftharpoons CO(g) + H_2O(g) \quad \Delta H_r^0 = +41.2\,kJ/kmol \tag{7.6}$$

Alkaline electrolysis, which uses an aqueous alkaline solution (KOH or NaOH) as the electrolyte, is currently the cheapest technology; however, in the future, polymer electrolyte membrane (PEM) electrolysis could be better suited for the PtG process chain [29]. The reasons for this are that PEM has a faster cold start, is more flexible, and has better coupling with dynamic and intermittent systems. The hydrogen produced in the electrolysis reaction cannot be directly injected into a natural gas grid as there are limits on the quantity of hydrogen allowed. For an example on these limits, from Holland see Table 6.3. The main drawbacks nowadays of Power to Gas are a relatively low efficiency and high costs but both are expected to improve in the future.

Another concept is to add the hydrogen directly into an anaerobic digester or into a separate reactor. This is referred to as biological methanation with microorganisms producing methane directly from CO_2 and H_2. Burkhardt and Busch [30] fed hydrogen and carbon dioxide into a patented reactor and using biological methanation achieved a final biogas with 97.9% methane. If this is scaled commercially there would be no need for upgrading techniques, just some basic gas cleaning operations would be needed. They also claimed that conventional biogas plants could be used with hydrogen addition to produce methane-enriched biogas.

7.4 Future Road Map

The authors suggest the following road map for the future development of biomethane technologies. The first stage is an R&D plan to improve the raw materials used in the production of biogas. Research into pretreatment methods can support various types of wastewater. Monitoring of the Carbon-to-Nitrogen ratio (C/N ratio) can be done in order to increase biogas yield. The optimal ratio is believed to be 25–30 parts carbon to 1 part nitrogen; however, the optimal should be found for each type of wastewater. The goal should be to increase the yield by 20%.

Research should be carried out to improve the efficiency of the biogas production process. Novel microbial molecular techniques can be targeted to increase the rate of biogas production or increase the purity of biogas. Future biogas production systems should be designed to support high organic loading rates (OLR), reducing construction and operating costs. Conduct research to find the remaining biogas potential in agricultural waste such as cassava pulp, cow manure, rice straw, sugarcane leaves, organic waste, and others. The focus should be on overall systems implementation from transportation, logistics, sorting process, material preparation, and pretreatment all with the end goal to increase the rate of biogas production.

It should be possible to implement a 20% cost reduction target on biogas production. Processes such as dry fermentation, complete stirring fermentation (CSTR), increase the fermentation efficiency by increasing the temperature or pressure. Techniques should be developed for the efficient digestion of cellulose (lignocellulosic digestion). The development of intelligent control devices and designing low-cost

Table 7.4 R&D pathways for biogas production

Biogas from waste		Biogas from energy crops	
Agricultural and municipal wastes		Energy crops	
		Napier grass	Advanced energy crops
Upstream	Poultry/cassava pulp Rice straw/cane leaves, etc. Harvesting machines Feedstock pretreatment Municipal waste sorting	Market study/logistics Plantation zoning yield Improvement harvesting Machine silage making technique	Algae Screening/GMO Algae cultivation/harvest Yeast cultivation/harvest
Processing	Dry fermentation CSTR or digester design Low-cost mixing Increase OLR Thermophilic/high Pressure smart control	Napier grass refinery Valuable product extraction Lignocellulosic digestion Reactor design Mixed waste digestion	Biorefinery reactor design Increase OLR
Downstream	Grid stability hybrid with other renewable H2S Scrubber/moisture Removal/flare	Digestate utilization Nutrient recycling	Zero-waste approach

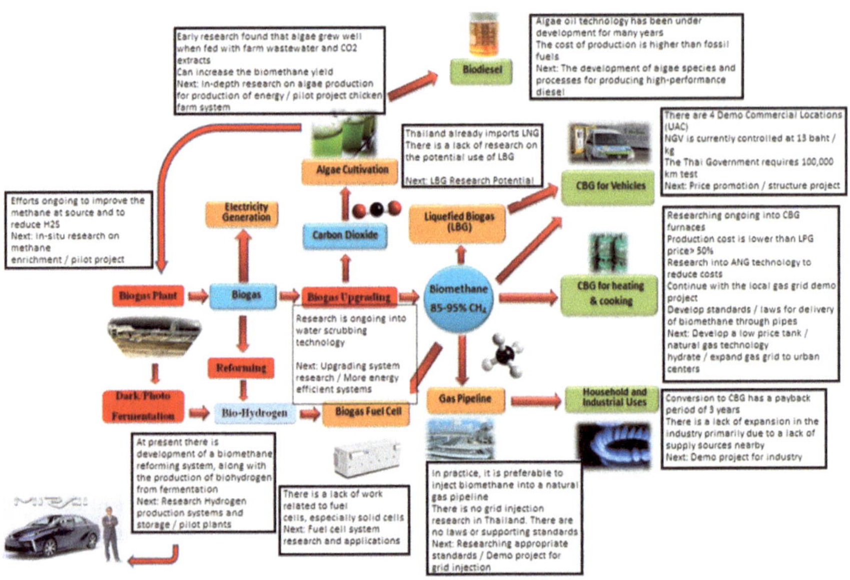

Fig. 7.5 ERDI's road map for biomethane research and development

Table 7.5 R&D pathways for biomethane production

Biomethane		
Biomethane plant		
Upstream Co-digestion Hydrolysis technique C/N balance	Upgrading technique	Biomethane use diversification
Processing Biorefinery Microbial technique Smart monitoring Increase OLR Low-cost digester Yield enhancement	Upgrading efficiency Upgrading cost reduction Compressor fabrication Membrane fabrication In situ CH_4 enrichment Cryogenics Water scrubbing Steam reforming New upgrading tech Advanced methane storage	Vehicle – CBG market study – Vehicle testing Pipeline – Local biogas grid – National grid injection – Gas pool mechanism Heat – Household – Industries CO_2 utilization/trading – algae growth enhance – power to gas – carbon credit trading Biohydrogen – production/storage Direct methane fuel cell – efficiency Gas-to-liquid fuel – Fischer–Tropsch
Downstream waste heat utilization Genset improvement Burner improvement		

production equipment will be done. As will improvements in gas compressors for high efficiency and low operating costs.

Conduct research on the cost and market mechanisms of nonfood or energy crops such as Napier grass. Planting technology to increase the yield per acre of energy crops. Large-scale harvesting machine development and implementation. Storing and preserving energy crops effectively. Genetic engineering of new energy crops such as algae and yeast to boost biogas yields.

After the production process, the efficient use of biomethane should be promoted. Efficient H_2S and moisture removal systems should be promoted. Research and development of biogas quality improvement technology for higher efficiency and less energy consumption. Development of quality measurement and analysis tools for production. Small-scale, efficient, distributed biomethane production systems should be encouraged using low-cost new techniques, such as in situ methane enrichment, improving gas quality with algae, improving water, chemical scrubbing, and pressure swing absorption systems. Waste heat from generators should be used wherever possible. Efficient flare systems and high-efficiency burners will increase biomethane utilization. The future stability of power generation systems that include renewables needs to be improved. Biomethane systems should be integrated with other energy sources such as solar and wind. In the future a demonstration of an economically feasible large scale biomethane upgrading plant will be completed. Research to utilize any remaining waste from digestate management, nutrient recycling, and zero-waste management policies should be carried out.

Conduct market research and guidelines to promote the production of CBG as a vehicle fuel and analyze the long-term performance of fuel in real engines. The study of the pollutants from CBG combustion must be conducted. Conduct research on biomethane gas transmission networks for low cost and high security, primarily for use in communities and industries as local gas grids. Improvement of household equipment such as cooking stoves and water heaters can be done. Air conditioners can be developed that can run off biomethane. The Gas Pool Price Mechanism can be studied to promote the use of CBG.

Research on the utilization of carbon dioxide which is separated from biogas, such as accelerating the growth of algae and trees and use of CO_2 as a raw material for the synthesis of methane gas. The development of a cryogenics system to dehumidify the biomethane and research on hydrogen production using steam reforming. Development of low-pressure methane and hydrogen storage systems and technologies for converting biomethane gas into liquid hydrocarbons (gas to liquid). These targets and goals are shown in Tables 7.4 and 7.5 along with the biogas development plan from the Energy Research and Development Institute Nakornping in Fig. 7.5.

References

1. Speirs J, Balcombe P, Johnson E, Martina J, Brandon N, Hawkes A (2018) A greener gas grid: what are the options. Energy Policy 118:291–297
2. Grando RL, de Souza M, Antune A, ValÃcria da Fonseca F, SÃinchez A, Barrena R, Font X, (2017) Technology overview of biogas production in anaerobic digestion plants: a european evaluation of research and development. Renew Sustain Energy Rev 80:44–53
3. COWI (2015) Study on actual ghg data for diesel, petrol, kerosene and natural gas. Technical report ENER/C2/2013-643, European Commission
4. Herrero Garcia N, Mattioli A, Gil A, Frison N, Battista F, Bolzonella D (2019) Evaluation of the methane potential of different agricultural and food processing substrates for improved biogas production in rural areas. Renew Sustain Energy Rev 112:1–10
5. Kovacs A (2013) Proposal for a European biomethane roadmap. Technical report, Green Gas Grids. www.greengasgrids.eu
6. EBA (2016) Biomethane in transport. Technical report, European Biogas Association, Rue Arlon 63-65 1040 Brussels Belgium
7. Mao C, Feng Y, Wang X, Ren G (2015) Review on research achievements of biogas from anaerobic digestion. Renew Sustain Energy Rev 45:540–555
8. Lee DH, Behera SK, Kim JW, Park HS (2009) Methane production potential of leachate generated from korean food waste recycling facilities: a lab-scale study. Waste Manag 29:876–882
9. Kim J, Park C, Kim TH, Lee M, Kim S, Kim SW (2003) Effects of various pretreatments for enhanced anaerobic digestion with waste activated sludge. J Biosci Bioeng 95:271–275

10. Nghiem DL, Wickham R, Ohandja DG (2017) Enhanced biogas production and performance assessment of a fullscale anaerobic digester with acid phase digestion. Int Biodeterior Biodegrad 124:162–168
11. Xia A, Cheng J, Murphy JD (2016) Innovation in biological production and upgrading of methane and hydrogen for use as gaseous transport biofuel. Biotechnol Adv 35(5):451–472
12. Saratale RG, Kumar G, Banu R, Xia X, Periyasamy S, Saratale GD (2018) A critical review on anaerobic digestion of microalgae and macroalgae and co-digestion of biomass for enhanced methane generation. Bioresour Technol 262:319–332
13. Wu N, Moreira MC, Zhang Y, Doan N, Yang S, Phlips JE, Svoronos AS, Pullammanappallil CP (2019) Anaerobic digestion, intechopen, chap techno-economic analysis of biogas production from microalgae through anaerobic digestion, pp 1–33. https://doi.org/10.5772/intechopen.86090
14. Xiaoqiang W, Nordlander E, Thorin E, Jinyue Y (2013) Microalgal biomethane production integrated with an existing biogas plant: a case study in Sweden. Appl Energy 112:478–484
15. Amirta R, Tanabe T, Watanabe T, Honda Y, Kuwahara M, Watanabe T (2006) Methane fermentation of japanese cedar wood pretreated with a white rot fungus, ceriporiopsis subvermispora. J Biotechnol 123:71–77
16. Zhao J (2013) Enhancement of methane production from solid-state anaerobic digestion of yard trimmings by biological pretreatment. Master's thesis, Ohio State University
17. Zhong W, Zhang Z, Luo Y, Sun S, Qiao W, Xiao M (2011) Effect of biological pretreatments in enhancing corn straw biogas production. Bioresour Technol 102:11177–11182
18. Mosier N, Wyman C, Dale B, Elander R, Lee Y, Mea Holtzapple (2005) Features of promising technologies for pretreatment of lignocellulosic biomass. Bioresour Technol 96:86–673
19. Rao P, Seenayya G (1994) Improvement of methanogenesis from cow dung and poultry litter waste digesters by addition of iron. World J Microbiol Biotechnol 10:211–214
20. Peterssen A, Wellinger A (2009) Biogas upgrading technologies—developments and innovations. Technical report, IEA Bioenergy
21. Linberg A (2003) Development of in-situ methane enrichment as a method for upgrading biogas to vehicle fuel standard. Master's thesis, Stockholm, KTH
22. Singhal S, Agarwal S, Arora S, Sharma P, Singhal N (2017) Upgrading techniques for transformation of biogas to bio-CNG: a review. Int J Energy Res. https://doi.org/10.1002/er.3719
23. Sahota S, Shaha G, Ghosha P, Kapoor R, Sengupta S, Singh P, Vijay V, Sahay A, Vijay V, Thakur I (2018) Review of trends in biogas upgradation technologies and future perspectives. Bioresour Technol Rep 1:79–88
24. Sirichai K, James M, Pruk A, Asira B (2018) Low pressure biomethane gas adsorption by activated carbon. Energy Sustain Dev 43:196–202
25. Chiew Y, Brown M, You V, Judd R, Briggs I (2011) Alternatives to venting of natural gas. In: International gas union research conference, UK
26. Burchell T, Rogers M (2000) Low pressure storage of natural gas for vehicular applications. SAE Trans 109(4):2242–2246
27. Pellegrini L, De Guido G, Lang S (2018) Biogas to liquefied biomethane via cryogenic upgrading technologies. Renew Energy 124:75–83
28. Tao L, Lin CY, Dou S, Feng S, Chen D, Liu D, Huo J, Xia Z, Wang S (2017) Creating coordinatively unsaturated metal sites in metal-organic-frameworks as efficient electrocatalysts for the oxygen evolution reaction: Insights into the active centers. Nano Energy 41:417–425
29. Gotz M, Lefebvre J, Mors F, McDaniel Koch A, Graf F, Bajohr S, Reimert R, Kolb T (2016) Renewable power-to-gas: a technological and economic review. Renew Energy 85:1371–1390
30. Burkhardt M, Busch G (2013) Methanation of hydrogen and carbon dioxide. Appl Energy 111:74–79

Index

A
Abnormal operations, 153
Acetogenesis, 1, 164
Acidogenesis, 1, 164
Activated carbon, 54
Activated carbon method, 17
Adjustable nozzle, 113
Adjustment valve, 119
Adsorbed natural gas, 169
Amine solvent, 57
Ammonia, 19
Anaerobic process, 1

B
Barlows formula, 148
Barrer, 63
Bernoulli, 116, 126
Biofilter, 51
Biological activated carbon, 23
Biological methanation, 175
Biological removal method, 17
Biomethane fuel delivery system, 130
Blowoff, 118
Burner head, 113

C
Calcium hydroxide, 58
Capacity factor, 64
Capital cost, 131
Carbon dioxide, 1
Carbon molecular sieve, 18, 54
Carbon-to-nitrogen ratio, 175
Carbonyl sulfide, 17
Cassava pulp, 175
Catering equipment, 123
Ceramic, 124
Ceramic kiln, 124
Chemical Oxygen Demand (COD), 3
Chemical scrubbing, 47, 57
Chiller, 72
Chlorine, 54
Chlorine dioxide, 54
Class location, 145
Class location unit, 145
Coefficient of performance, 41
Combustion, 121
Complete stirring fermentation, 175
Compressed Biomethane Gas (CBG), 4
Compressed Natural Gas (CNG), 4, 115
Condenser, 39
Corrosion, 151

D
Dedicated CNG, 103
Desiccant, 73
Design factor, 146
Desorber, 50
Dew point, 72
Dew point temperature, 19, 40
Diesel dual fuel, 104
Diethanolamine, 57
Diglycolamine, 57
Domestic cooking, 111
Domestic stoves, 111
Drain valve, 158
Dual pressure low-temperature distillation, 173

E
Economic analysis, 64

S. Koonaphapdeelert et al., *Biomethane*, Green Energy and Technology,
https://doi.org/10.1007/978-981-13-8307-6

Economic potential, 111
Emergency plans, 153
Emissions, 129
Energy conservation, 111
Epoxy resin, 112
Equilibrium capacity, 74
Ethylene glycol, 76
Ethylene mercaptan, 82
Evaporator, 40
Excess air, 121
Excess flow valve, 158
Exothermic, 50
Extinguish, 121

F
Fiber tank, 112
Ficks law, 62
Fittings, 145
Flanges, 145
Flashback, 117
Fossil fuels, 117
Fuel LHV, 113

G
Gate fees, 69
Glass fiber, 112
Granular activated carbon, 23
Greenhouse gas emissions, 9
Greenware, 133

H
Halogenated hydrocarbons, 20
Hazardous leak, 151
Heating time, 123
Henrys law, 49
High-performance polyimides, 63
Hoop stress, 148
Humidity, 18
Hydraulic diameter, 86
Hydraulic retention time, 165
Hydrogen peroxide, 54
Hydrogen sulfide, 1, 17
Hydrolysis, 1, 164
Hygroscopic, 74

I
Ideal gas law, 169
Industrial gas separation, 59
Industrial price of LPG, 130
Injection nozzle, 127
Inspection, 154
Intake manifold, 118
Intelligent control devices, 177
Interchangeability, 115
Internal rate of return, 134, 160
Iron Chloride Dosing Method, 17
Iron oxide pellets method, 17
Iron sponge method, 17

L
Laminar burning velocity, 117
Lampang, 133
Leakage history, 152
Lignocellulosic biomass, 166
Liquefied biomethane, 172
Liquid biomethane, 167
Longitudinal joint factor, 147
LPG, 121
LPG subsidy, 111

M
Mass transfer zone, 74
Mee Sin Ceramics, 133
Membrane separation, 47, 61
Metal-organic framework, 169
Meter, 150
Methane, 1, 4
Methane content, 112
Methane number, 103
Methane slip, 62
Methanogenesis, 1, 164
Microbubble generator, 58
Microorganism, 1, 54
Molar flow rate, 118
Molecular sieve, 76
Monoethanolamine, 57, 58
Mueller equation, 157

N
Napier grass, 177
Natural Gas for Vehicles (NGV), 4
Net positive suction head, 86
Nitrogen, 1, 20
Nitrogen oxides, 19

O
Odorization, 82
Operating costs, 160
Operating pressure, 158
Optimal, 121, 129

Organic loading rate, 164, 165, 175
Oxygen, 1
Oxygen dosing method, 17

P
Parameters, 121
Particulates, 19
Payback period, 124, 134
Peracetic acid, 54
Photobioreactor, 166
Physical solvent, 59
Pollutants, 129
Polymer electrolyte membrane, 175
Polymers of intrinsic microporosity, 63
Power-to-gas, 174
Premixing section, 117
Pressure drop, 120
Pressure swing adsorption, 47, 54
Programmable logic controller, 92
Purity, 140

R
Radial burner, 112, 124
Reactants velocity, 117
Refrigerant, 40
Refuse derived fuel, 12
Regulator, 150
Renewable energy, 5
Renewable Natural Gas (RNG), 4
Restrictor, 119
Retention time, 35
Rocket burner, 128
Room-temperature ionic liquids, 63

S
Scale, 123
Selexol process, 59
Shutoff valve, 158
Shuttle kiln, 127, 134
Silica gel, 75
Siloxanes, 20
Sintering, 124
Sodium hydroxide, 58
Standard dimensional ratio, 158
Sugarcane leaves, 175
Sulfide-oxidizing bacteria, 24
Swirl burner, 112, 124

T
Temperature derating factor, 147
Temperature swing absorption, 47
Temperature swing adsorption, 56
Test medium, 151
Tetrahydrotiophen, 82
Thermal expansion, 149
Thermally rearranged polymers, 63
Thermocouple, 128
Thermodynamic properties, 118
Thermoplastic pipe, 148
Thermosetting plastic, 148
Tons of oil equivalent, 7

U
Underground clearance, 149

V
Virgin activated carbon, 23
Volatile fatty acids, 165
Volatile solids, 165

W
Waste activated sludge, 165
Water scrubbing, 47
Water scrubbing method, 17
Wobbe index, 115, 116, 141

Z
Zeolites, 54
Zeolitic imidazolate frameworks, 169

MIX
Papier aus verantwortungsvollen Quellen
Paper from responsible sources
FSC® C105338
FSC
www.fsc.org

If you have any concerns about our products,
you can contact us on
ProductSafety@springernature.com

In case Publisher is established outside the EU,
the EU authorized representative is:
Springer Nature Customer Service Center GmbH
Europaplatz 3, 69115 Heidelberg, Germany

Printed by Libri Plureos GmbH
in Hamburg, Germany